Wireless Data Technologies

Wireless Data Technologies

Vern A. Dubendorf

IEEE, Massachusetts Health Data Consortium (MHDC), Mobile Healthcare Alliance (MOHCA) and The Council of Communications Advisors USA

WILEY

Other Wiley Editorial Offices

John Wiley & Sons Inc., 111 River Street, Hoboken, NJ 07030, USA

Jossey-Bass, 989 Market Street, San Francisco, CA 94103-1741, USA

Wiley-VCH Verlag GmbH, Boschstr. 12, D-69469 Weinheim, Germany

John Wiley & Sons Australia Ltd, 33 Park Road, Milton, Queensland 4064, Australia

John Wiley & Sons (Asia) Pte Ltd, 2 Clementi Loop #02-01, Jin Xing Distripark, Singapore 129809

John Wiley & Sons Canada Ltd, 22 Worcester Road, Etobicoke, Ontario, Canada M9W 1L1

Wiley also publishes its books in a variety of electronic formats. Some content that appears
in print may not be available in electronic books.

Library of Congress Cataloging-in-Publication Data

Dubendorf, Vern A.
  Wireless data technologies / Vern A. Dubendorf.
      p. cm.
  Includes bibliographical references and index.
  ISBN 0-470-84949-5 (alk. paper)
    1. Wireless communication systems – handbooks, manuals, etc. 2. Data transmission
systems – Handbooks, manuals, etc. I. Title.

  TK5103.2 .D83 2003
  621.382 – dc21

British Library Cataloguing in Publication Data

A catalogue record for this book is available from the British Library

ISBN 0-470-84949-5

Typeset in 11/13pt Times by Laserwords Private Limited, Chennai, India
Printed and bound in Great Britain by Antony Rowe Ltd, Chippenham, Wiltshire
This book is printed on acid-free paper responsibly manufactured from sustainable forestry
in which at least two trees are planted for each one used for paper production.

In the Beginning, ARPA created the ARPANET.

And the ARPANET was without form and void.

And darkness was upon the deep.

And the spirit of ARPA moved upon the face of the network and ARPA said, 'Let there be a protocol,' and there was a protocol. And ARPA saw that it was good.

And ARPA said, 'Let there be more protocols,' and it was so. And ARPA saw that it was good.

And ARPA said, 'Let there be more networks,' and it was so.

Danny Cohen

Contents

Dedication

I would like to dedicate this work to three very special people who have had a profound impact in my life and in the creation of this work.

To Dana Axinger, whose quest for knowledge was the greatest driving factor towards the development of this work.

To my wonderful wife Louann who is becoming a 'Wireless Guru' in her own right by having spent numerous hours proof reading this work and suggesting methods of explaining ideas so that they are easy to comprehend. She has also provided input from her training and knowledge of Ultrasound Technology, which is also in the realm of Wireless Technologies.

To my mother-in-law, Ellamarie Carr, whose background in education helped lead me to the desire to complete this work.

Foreword

There are few occasions in life where one gets the opportunity to work with someone that actually blazes new trails. Working with Dr Tony Dubendorf over the past couple of years has afforded me an opportunity to understand the world of wireless communications better than my previous 20-odd years of experience in the IT and networking world. Tony exhibits more than just a grasp and mastery of the material related to wireless communications and wireless security. He exudes wireless in every professional act that he performs. His thoughts are always centered on how to improve communications or data security or packet transmission or something related to wireless communications. I admit that I have had a serious conflict in my career between work and fun – being in IT for so long does that to folks. However, I have not even approached the level of understanding in wireless communications that Tony exhibits. Now, to see this information put in a book that is well constructed, well written, and full of information related to wireless history and wireless technological developments of late, I am excited to recommend this book to anyone interested in learning about the new medium of communications for the future.

It is always good when someone with such expertise attempts to convey what he or she knows to others in order to benefit the greater society of technologists. Tony has done just that. He has written a book that takes the reader from the very fundamental aspects of the history of networking and wireless to the latest advances in the field in a gradual, methodical progression of building blocks. This approach allows the reader to stay focused on each particular topic and understand what is read without having to refer back to previous sections for clarification. His masterful explanations make short work of these topics and I can only say that the book is much better than most textbooks I have come across in my time.

For any student or technologist wanting to get a clear, concise, and well-rounded explanation of what wireless means in today's communications world, I would heartily recommend reading this book. It is destined to be a 'shelf-reference' we tech folks keep handy. You can always tell by the dog-eared pages how useful a book is to the reader and I suspect this book will have many such 'dog-ears' as it becomes mainstream in the technical marketplace.

John Rittinghouse, Ph.D.
Managing Partner
Aphanes ProServe, LLC

Preface

While many industry experts around the world have dubbed me as a 'Wireless Guru', I find the need to provide my definition of 'Wireless Guru'. A 'Wireless Guru' is someone who has great knowledge of the entire range of wireless technologies while having an expertise in at least one field. My fields of choice to obtain expertise are in Wireless Data Networks and Wireless Information Security. In this book I have endeavored to provide you with a basic understanding of each prevalent wireless technology and Information Security.

I trace back my interest in wireless technology to the 1960s when I built my first Heathkit Radio. In the 1970s I expanded into Amateur Radio, Packet Data, and the study of Meteorology. Today I design wireless access to information networks and network infrastructures. I am also involved in the development of Wireless Information Security techniques and policies.

Wireless Technologies in general have been growing at a rate that out-shadows the public advent of the Internet. In this book, I use the term 'wireless technologies' to include any type of voice or data communications that uses the 'ether' as the transmission medium.

The mobile Internet is rapidly taking off. Fueling the increased interest in Wireless Local Area Networks (WLANs) is the relative and growing ease of availability of wireless Internet access for notebook computers, handhelds such as the Palm and iPAQs and the increasingly common use of wireless networks in the home and office. You can now have a wireless Internet connection while you are traveling almost anywhere in the United States. Airports, hotels, coffee shops and convention centers are quickly installing WLANs.

A recent development in wireless Internet access is the ongoing deployment of Pirate Wireless Networks. These are groups of people that join together to install wireless Local Area Networks (LANs), Wide Area Networks (WANs) and Metropolitan Area Networks (MANs) around the globe, which are normally accessible at no cost, providing wireless access to the Internet for end users. Pirate Wireless Networks can be found from San Francisco to Boston in the US, London, Frankfurt, Brisbane, Tokyo, Sidney, Taipei, Glasgow, Dublin, Moscow, and almost any other major metropolitan area on the face of the earth.

To take advantage of the tremendous growth rate that is being experienced in Wireless Technologies, service providers have been fiercely competing to increase market share and build a customer base, while manufacturers such as Nokia, Eriksson, and Motorola are creating new user interfaces and feature sets on their telephones enabling Mobile Internet access. The anticipated consumer demand for high-bandwidth wireless data is commonly seen as the driving force behind current network upgrades and expansions. The type of companies investing in wireless technologies illustrates the importance of wireless data. Nontraditional telecommunications companies such as Cisco Systems, Microsoft, Intel, 3Comm, Compaq, Hewlett-Packard and many others, are investing heavily in wireless product development in both hardware and software.

Now we have the dawning of an evolution in Information Security (InfoSec) that must envelop the emerging mobile information access and wireless technologies. We are starting to see a shift from the traditional wireless Internet access to the more demanding mobile information access. This is occurring across all sectors of business such as financial institutions, medical facilities and practices, and corporate networks and is evident by the development of HIPAA, which is the abbreviation for the Health Insurance Portability & Accountability Act of 1996 (August 21), Public Law 104–191, which amends the Internal Revenue Service Code of 1986, also known as the Kennedy–Kassebaum Act, and other security standards.

Thus, wireless technology is the next information age evolution. With that in mind, wireless technology must be a fundamental part of whatever system you are designing, whether it's a network infrastructure, software package, or database access.

In this book, I'll examine many of these wireless systems and information securities. I'll also attempt to look into the future and see what's in store for Wireless Technologies.

Acknowledgments

This book would not have been possible without the help and encouragement of many people. I would like to thank them for their efforts in improving the end results. However, before I do, I need to mention that I have done my best to correct mistakes that the reviewers have pointed out. I alone am responsible for any remaining errors.

I would like to begin by giving special thanks to Dr William (Bill) Hancock of Exodus, a Cable & Wireless Service, for his expertise in network and information security, Dr John Rittinghouse of Aphanes, Inc. for his desire to learn more about wireless technologies and wireless information security, James Ransome of Exodus, a Cable & Wireless Service, for his vast knowledge of Information Security and his continuing work on 'A Multilevel, Multirisk, Security Model for Next Generation Wireless Security Networks,' and Kevin Nixon of Exodus, a Cable & Wireless Service, for his expertise in Information Security in the Financial and Healthcare fields. All have spent many hours in reviewing early drafts of this book as well as providing ideas for additional information. Their comments and insights undoubtedly improved the quality of this work.

I would like to thank Bruce Schneier for his written works on Applied Cryptography that I find so fascinating and which have been an impetus in my quest for expertise in the field of Information Security.

I also need to thank Bob Antia for his vast knowledge of the Internet infrastructure and his willingness to share his knowledge and his love for the color orange.

Next I would like to thank Mark Hammond, Publisher and Sarah Hinton, Development Editor at John Wiley & Sons Ltd., West Sussex, England, for their patience and guidance throughout the course of this project.

I would like to thank Steve Marino. He is one of the true great minds that absorbs technology like a sponge and has the ability to use that knowledge to help his clients. Truly one of the great Strategic Account Executives.

I would like to thank Scott Prouty and Arbor Networks for taking me deep inside the world of 'Denial of Service' attacks and the great impact that it has and will continue to have on Wireless Technology.

Finally, I am grateful to Exodus, a Cable & Wireless Service, Lucent Technologies, Cisco Systems and many others for providing me with the opportunity to work in an exciting engineering environment on many rewarding projects.

1

A History of Wireless Technologies

1.1 Introduction

Radio or wireless–there must be a big difference; they are spelled a little different.

I hate to disappoint and disillusion some of you who have counted so much on a 'big difference'. However, just brace yourselves and prepare for the shock: *there is absolutely no difference between radio and wireless except the spelling.*

Wireless does not mean sparks, noise, or a lot of switches. Wireless means communication without the use of wires other than the antenna, the ether, and ground taking the place of wires. Radio means exactly the same thing: it is the same process. Communications by wireless waves may consist of an SOS or other messages from a ship at sea or the communication may be simply the reception of today's top 10 music artists, or connecting to the Internet to check your email. It does not become something different in either spelling or meaning. Table 1.1 demonstrates a simple timeline of Wireless Technologies evolution.

1.2 Where it all began – Marconi

In February 1896, Guglielmo Marconi journeyed from Italy to England in order to show the British telegraph authorities what he had developed in the way of an *operational* wireless telegraph apparatus. His first British patent application was filed on June 2 of that year.

Through the cooperation of Mr W.H. Preece, who was at that time the chief electrical engineer of the British Post-office Telegraphs, signals were sent in July 1896 over a distance of one-and-three-fourths miles on Salisbury Plain.

Wireless Data Technologies. Vern A. Dubendorf
© 2003 John Wiley & Sons, Ltd ISBN: 0-470-84949-5

Table 1.1 A simple timeline in Wireless Technologies evolution (this is not to be considered an 'all inclusive' timeline)

1896	Guglielmo Marconi develops the first wireless telegraph system
1927	First commercial radiotelephone service operated between Britain and the US
1946	First car-based mobile telephone set up in St. Louis, using 'push-to-talk' technology
1948	Claude Shannon publishes two benchmark papers on Information Theory, containing the basis for data compression (source encoding) and error detection and correction (channel encoding)
1950	TD-2, the first terrestrial microwave telecommunication system, installed to support 2400 telephone circuits
1950s	Late in the decade, several 'push-to-talk' mobile systems established in big cities for CB-radio, taxis, police, etc.
1950s	Late in the decade, the first paging access control equipment (PACE) paging systems established
1960s	Early in the decade, the Improved Mobile Telephone System (IMTS) developed with simultaneous transmit and receive, more channels, and greater power
1962	The first communication satellite, Telstar, launched into orbit
1964	The International Telecommunications Satellite Consortium (INTELSAT) established, and in 1965 launches the Early Bird geostationary satellite
1968	Defense Advanced Research Projects Agency – US (DARPA) selected BBN to develop the Advanced Research Projects Agency Network (ARPANET), the father of the modern Internet
1970s	Packet switching emerges as an efficient means of data communications, with the X.25 standard emerging late in the decade

Table 1.1 (*continued*)

1977	The Advanced Mobile Phone System (AMPS), invented by Bell Labs, first installed in the US with geographic regions divided into 'cells' (i.e. cellular telephone)
1983	January 1, TCP/IP selected as the official protocol for the ARPANET, leading to rapid growth
1990	Motorola files FCC application for permission to launch 77 (revised down to 66) low earth orbit communication satellites, known as the Iridium System (element 77 is Iridium)
1992	One-millionth host connected to the Internet, with the size now approximately doubling every year
1993	Internet Protocol version 4 (IPv4) established for reliable transmission over the Internet in conjunction with the Transport Control Protocol (TCP)
1994–5	FCC licenses the Personal Communication Services (PCS) spectrum (1.7 to 2.3 GHz) for $7.7 billion
1998	Ericsson, IBM, Intel, Nokia, and Toshiba announce they will join to develop Bluetooth for wireless data exchange between handheld computers or cellular phones and stationary computers
1990s	Late in the decade, Virtual Private Networks (VPNs) based on the Layer 2 Tunneling Protocol (L2TP) and IPSEC security techniques become available
2000	802.11(b)-based networks are in popular demand
2000–1	Wired Equivalent Privacy (WEP) Security is broken. The search for greater security for 802.11(x)-based networks increases

In March 1897, a greater distance of four miles on Salisbury Plain was covered with wireless signals. On May 13 of that same year, communication was established between Lavernock Point and Brean Down England, at distance of eight miles.

In America, during the period of 1890 to 1896, many students of science were in touch with the discoveries made in Europe during this period; but it was not until 1897 that the utilitarian American mind sensed the commercial possibilities of the advances being made abroad.

In its March 1897 issue, *McClure's Magazine* presented a long illustrated article entitled 'Telegraphing Without Wires,' by H.J.W. Dam, describing the experiments of Hertz, Dr Chunder Bose, and the youthful Marconi.

Telegraph Age, New York, in its issues of November 1 and November 15, 1897, reprinted a long article from the London *Electrician*, entitled 'Marconi Telegraphy.' This article consisted chiefly of the technical description that accompanied Marconi's British patent specification number 12 039 of 1896.

In September 1899, during the International Yacht Races held off of New York harbor, the steamer *Ponce* was equipped with radio devices by Marconi, for the purpose of transmitting reports on the progress of the race. Two receiving stations were equipped: one on the Commercial Cable Company's cable ship *Mackay Bennett*, stationed near Sandy Hook, and connected with a land line station on shore by means of a regulation cable; the other at Navasink Highlands. This demonstration, even though it wasn't very successful, immediately brought the subject to the front in the United States interest. In 1900, the erection of the first Marconi station at Cape Cod, Massachusetts, began.

In March 1901, the Marconi Company installed radio devices at five stations on five islands of the Hawaiian group. For a long time these installations were to prove to be of little or no value due to the restricted availability scarcity of qualified operatives.

During this same year, the Canadian government installed two stations in the Strait of Belle Isle; also constructed were the New York *Herald* stations at Nantucket, MA, and Nantucket light ship.

The greatest radio event of 1901 was the reception by Dr Marconi at St Johns, Newfoundland, of what has become known as the famous letter 'S', transmitted as a test signal from his English station; this was on December 11, 1901.

In 1904, several US government agencies, which included the Navy, the Department of Agriculture, and the Army's Signal Corps, all began setting up their own radio transmitters, with little or no coordination between the various departments. In 1904, President Theodore Roosevelt appointed a board, which consisted of representatives from these agencies. This board was tasked with preparing recommendations for coordination of governmental development of radio services. The 1904 'Roosevelt Board' Report proposed assigning most of the oversight of government radio to the Navy Department and proposed imposing significant restrictions on commercial stations.

1.3 Packet Data

Packet Data technology was developed in the mid-1960s and was put into practical application in the ARPANET, which was established in 1969. Initiated in 1970, the ALOHANET, based at the University of Hawaii, was the first large-scale packet radio project.

Amateur packet radio began in Montreal, Canada, in 1978 with the first transmission occurring on May 31. This was followed by the Vancouver Amateur Digital Communication Group (VADCG) development of a Terminal Node Controller (TNC) in 1980.

The current TNC standard grew from a discussion in October of 1981 at a meeting of the Tucson Chapter of the IEEE Computer Society. A week later, six of the attendees gathered together and discussed the feasibility of developing a TNC that would be available to amateurs at a modest cost. The Tucson Amateur Packet Radio Corporation (TAPR) was formed from this project. On June 26 1982, Lyle Johnson and Den Connors initiated a packet contact with the first TAPR unit. The project progressed from these first prototype units to the TNC-1 and then finally to the TNC-2 which is now the basis for most packet operations worldwide.

Packet has three great advantages over other digital modes: transparency, error correction, and automatic control.

The operation of a packet station is transparent to the end user. Connect to the other station, type in your message, and it is sent automatically. The Terminal Node Controller (TNC) automatically divides the message into packets, keys the transmitter, and then sends the packets. While receiving packets, the TNC automatically decodes, checks for errors, and displays the received messages.

Packet radio provides error-free communications due to the built-in error detection schemes. If a packet is received, it is checked for errors and will be displayed only if it is correct. In addition, any packet TNC can be used as a packet relay station, sometimes called a digipeater. This allows for greater range by stringing several packet stations together.

Users can connect to their friends' TNCs at any time they wish, to see if they are at home. Another advantage of packet over other modes is the ability for many users to be able to use the same frequency channel simultaneously.

Since packet radio is most commonly used at the higher radio frequencies (VHF), the range of the transmission is somewhat limited. Generally, transmission range is limited to 'unobstructed line-of-sight' plus approximately 10–15% additional distance.

The transmission range is influenced by the transmitter power and the type and location of the antenna, as well as the actual frequency used and the length of the antenna feed line (the cable connecting the radio to the antenna).

Another factor influencing the transmission range is the existence of obstructions (hills, groups of buildings, etc.). Connections made in the 144–148 Mhz range could be 10 to 100 miles, depending on the specific combination of the variables mentioned above.

1.4 Voice Technologies

In the November 7, 1920 issue of the Boston *Sunday Post* there was an article authored by John T. Brady covering the topic of 'Talking by Wireless as You Travel by Train or Motor,' which noted 'It is now possible for a business man to talk with his office from a moving vehicle.' This was a review of two-way radio conversation tested by Mr Brady and with Harold J. Power who was then the head of the American Radio and Research Corporation, while Power was in a moving automobile.

It would not be until the 1980s that the technology needed for such things as pagers and wireless telephones would be perfected to the point that they became widely available consumer products. Although the telephone's use for individual communication largely overshadowed applications for distributing entertainment and news, the reverse would be true for radio, with broadcasting dominating for decades, before radio transmissions would be significantly developed for personal, mobile communication.

1.5 Cellular Technologies

In cellular networks there are radio ports with antennas that connect to base stations (BSs) that serve the user equipment known as mobile stations (MSs). The communication that takes place from the MS to the BS is knows as the uplink while the communication from the BS to the MS is known as the downlink. The downlink is contentionless, however several MSs access the uplink simultaneously. This uplink uses a very important characteristic, which is the multiple-access technique.

Frequency-division multiple access (FDMA), time-division multiple access (TDMA) and code-division multiple access (CDMA) are the most widely used physical-layer multiple access techniques in use today.

The infrastructures of cellular networks include mobile switching centers (MSCs). These control one or more BSs and provide the interface for them to the wired public switched telephone network (PSTN), a central home location register (HLR) and the visiting location register (VLR) for each MSC. The VLR and HLR are databases that keep the registered and current locations of MSs to be used in the handoffs. Handoff is the process of handing a call from one cell to a new cell as the MS moves around.

2

Understanding Spread Spectrum Technologies

2.1 Introduction

Spread Spectrum (SS) dates back to World War II. The allies also experimented with spread spectrum in World War II. These early research and development efforts tried to provide countermeasures for radar, navigation beacons, and communications. The US Military has used SS signals over satellites for at least 25 years. An old, but faithful, highly capable design like the Magnavox USC-28 modem is an example of this kind of equipment. Housed in two or three six-foot racks, it had selectable data rates from a few hundred bits per second to about 64 kbps and transmitted a spread bandwidth of 60 MHZ. Many newer commercial satellite systems are now converting to SS to increase channel capacity and reduce costs.

Wireless bridges using this technology are most commonly found operating in the 2.4 GHz unlicensed band. These provide data rates ranging from 1 to 11 Mbps creating an average rate of around 5.4 Mbps with distances up to 10 to 25 miles (16–40 km) depending on terrain, weather conditions, and the type of antenna used. The point to remember is that the greater the distance between the two points, the lower the throughput. Products in this category are also found operating in the 5.8 GHz (UNII) band. They employ either frequency hopping or direct sequence spread spectrum technologies.

The features offered by this technology are similar to those offered by wireline bridges:

- Interconnection with Ethernet or Token-Ring networks
- Spanning Tree Protocol

Wireless Data Technologies. Vern A. Dubendorf
© 2003 John Wiley & Sons, Ltd   ISBN: 0-470-84949-5

- Telnet for remote configuration
- SNMP (Simple Network Management Protocol)
- Automatic configuration using Bootstrap Protocol (BOOTP)
- File Transfer Protocol (FTP)
- DHCP (Dynamic Host Configuration Protocol)
- HTMP
- MIBs.

2.2 What Spread Spectrum Does

The use of these special pseudo noise codes in spread spectrum (SS) communications makes signals appear wide band and noise-like. It is this very characteristic that makes SS signals possess the quality of Low Probability of Intercept. SS signals are hard to detect on narrow band equipment because the signal's energy is spread over a bandwidth of maybe 100 times the information bandwidth.

The spread of energy over a wide band, or lower spectral power density, makes SS signals less likely to interfere with narrowband communications. Narrow band communications, conversely, cause little to no interference to SS systems because the correlation receiver effectively integrates over a very wide bandwidth to recover an SS signal. The correlator then 'spreads' out a narrow band interferer over the receiver's total detection bandwidth. Since the total integrated signal density or SNR at the correlator's input determines whether there will be interference or not. All SS systems have a threshold or tolerance level of interference beyond which useful communication ceases. This tolerance or threshold is related to the SS processing gain. Processing gain is essentially the ratio of the RF bandwidth to the information bandwidth.

A typical commercial direct sequence radio might have a processing gain of from 11 to 16 dB, depending on data rate. It can tolerate total jammer power levels of from 0 to 5 dB stronger than the desired signal. Yes, the system can work at negative SNR in the RF bandwidth, because of the processing gain of the receiver's correlator the system functions at positive SNR on the baseband data. Besides being hard to intercept and jam, spread spectrum signals are hard to exploit or spoof. Signal exploitation is the ability of an enemy (or a non-network member) to listen in to a network and use information from the network without being a valid network member or participant. Spoofing is the act of falsely or maliciously introducing misleading or false traffic or messages to a network. SS signals also are naturally more secure than narrowband radio communications. Thus, SS signals can be made to have any degree of message privacy that is desired. Messages can also be cryptographically encoded to any level of secrecy desired.

The very nature of SS allows military or intelligence levels of privacy and security to be accomplished with minimal complexity. While these characteristics may not be very important to everyday business and LAN (local area network) needs, these features are important to understand.

2.3 How Spread Spectrum Works

Spread Spectrum uses wide band, noise-like signals. Because Spread Spectrum signals are noise-like, they are hard to detect. Spread Spectrum signals are also hard to intercept or demodulate. Spread Spectrum signals are harder to jam (interfere with) than narrowband signals. These Low Probability of Intercept (LPI) and anti-jam (AJ) features are why the military has used Spread Spectrum for so many years. Spread signals are intentionally made to be much wider band than the information they are carrying to make them more noise-like.

Spread Spectrum transmitters use transmit power levels similar to narrow band transmitters. Because Spread Spectrum signals are so wide, they transmit at a much lower spectral power density, measured in Watts per Hertz, than narrowband transmitters. This lower transmitted power density characteristic gives spread signals a big plus. Spread and narrow band signals can occupy the same band, with little or no interference. This capability is the main reason for all the interest in Spread Spectrum today.

One way to look at spread spectrum is to understand that it trades a wider signal bandwidth for better signal-to-noise ratio. Frequency hopping and direct sequence are well-known techniques today. The following paragraphs will describe each of these common techniques in a little more detail and explain that pseudo noise code techniques provide the common thread through all spread spectrum types.

2.3.1 Frequency Hopping

Frequency hopping is the easiest spread spectrum modulation to use. Any radio with a digitally controlled frequency synthesizer can, theoretically, be converted to a frequency hopping radio. This conversion requires the addition of a pseudo noise (PN) code generator to select the frequencies for transmission or reception. Most hopping systems use uniform frequency hopping over a band of frequencies. This is not necessary if both the transmitter and receiver of the system know in advance what frequencies are to be skipped. Thus, a frequency hopper in two meters could be made to skip over commonly used repeater frequency pairs. A frequency-hopped system can use analog or digital carrier modulation and can be designed using conventional narrow band radio techniques. A synchronized pseudo noise code generator that drives the receiver's local oscillator frequency synthesizer does de-hopping in the receiver.

2.3.2 Direct Sequence

The most practical of all digital versions of SS is direct sequence (DSSS). A direct sequence system uses a locally generated pseudo noise code to encode digital data to be transmitted. The local code runs at a much higher rate than the data rate. Data for transmission is simply logically modulo-2 added (an EXOR operation) with the faster pseudo noise code. The composite pseudo noise and data can be passed through a data scrambler to randomize the output spectrum (and thereby remove discrete spectral lines). A direct sequence modulator is then used to 'double sideband suppressed carrier modulate' the carrier frequency to be transmitted. The resultant DSB suppressed carrier AM modulation can also be thought of as binary phase shift keying (BPSK). Carrier modulation other than BPSK is possible with direct sequence. However, binary phase shift keying is the simplest and most often used SS modulation technique.

An SS receiver uses a locally generated replica pseudo noise code and a receiver correlator to separate only the desired coded information from all possible signals. An SS correlator can be thought of as a very special matched filter – it responds only to signals that are encoded with a pseudo noise code that matches its own code. Thus, an SS correlator can be 'tuned' to different codes simply by changing its local code. This correlator does not respond to manmade, natural or artificial noise or interference. It responds only to SS signals with identical matched signal characteristics and encoded with the identical pseudo noise code.

2.4 Frequency Hopping Spread Spectrum

FHSS is a Frequency Modulation (FM) technique. FM modulates the frequency of the carrier wave with a modulating wave (Figure 2.1). The data is carried on a signal that jumps from one frequency to another in a preprogrammed, pseudo-random sequence. For the signal to be received correctly, the sequence must be known by the devices at both ends in advance. FHSS is probably more secure than DSSS since both receiver and transmitter must be in exact sequence, but it is also more expensive and complex to implement.

Figure 2.1   A spectrum analyzer photo of a Frequency Hop (FH) spread
spectrum signal

FHSS spreads the signal by transmitting a short burst on one frequency, 'hop-ping' to another frequency for another short burst and so on. The source and destination of a transmission must be synchronized so they are on the same frequency at the same time. The hopping pattern (frequencies and order in which they are used) and dwell time (time at each frequency) are restricted by most regulatory agencies. For example, the FCC requires that 75 or more frequencies be used and a maximum dwell time of 400 ms if interference occurs on one frequency, then the data is retransmitted on a subsequent hop on another frequency. All FHSS products on the market allow users to deploy more than one channel in the same area. This is accomplished by implementing separate channels on different, orthogonal, hopping sequences. Because there are a large number of possible sequences in the 2.4 GHz band, FHSS allows many non-overlapping channels to be deployed. FHSS systems have advantages over DSSS networks which include the following.

- FHSS is better at dealing with attenuation multipath interference (caused by the signal bouncing off walls, doors, or other objects and arriving at the destination at different times) by hopping to a different frequency that is not attenuated. The DSSS format is not capable of overcoming this effect due to the typical spreading factor used. DSSS does better if antenna diversity is used but building in antenna diversity causes products to be larger, heavier, and costlier.

- FHSS networks are able to provide three to four times more total network capacity than DSSS networks. In the 2.4 GHz band, the maximum number of non-overlapping 2M bps channels for a DSSS system is three (for a total of 6M bps capacity).
- Because of the nature of their synchronization, DSSS products do not permit roaming between channels. Roaming communities must be all on the same channel, thus creating a limit of one channel for most DSSS installations. If DSSS Access Points (APs) are placed on the same channel they will interfere with each other. If a new AP is placed on a different channel, users cannot roam to it.
- Topologies of large WLANs more or less guarantee that users will receive signals from multiple APs. This interference means that DSSS users will experience uneven performance depending on their physical location. FHSS networks allow users to place adjacent APs on different channels and avoid this problem. Even adjacent FHSS APs on the same channel will usually not interfere with each other. Although they share the same hopping sequence, they will usually not be synchronized in time. The result: they will rarely be at the same frequency at the same time. This is in contrast to the maximum of three DSSS networks that can overlap without constantly interfering with each other.
- Lightweight is critical for mobile applications. FHSS technology allows significantly lighter products to be developed. PCMCIA DSSS adapter cards are (typically) nearly twice as heavy as the equivalent for FHSS equivalents.
- Data from DSSS products is more easily intercepted than data from an FHSS product and, though DSSS and FHSS products can be supplemented with specialized encryption devices, doing so causes an increase in cost, weight, and power consumption. They also reduce performance by increasing round-trip delay.

Frequency hopping is the easiest spread spectrum modulation to use. Any radio with a digitally controlled frequency synthesizer can, theoretically, be converted to a frequency hopping radio. This conversion requires the addition of a pseudo noise (PN) code generator to select the frequencies for transmission or reception. Most hopping systems use uniform frequency hopping over a band of frequencies. This is not necessary if both the transmitter and receiver of the system know in advance what frequencies are to be skipped. Thus, a frequency hopper in two meters could be made that skipped over commonly used repeater frequency pairs. A frequency-hopped system can use analog or digital carrier modulation and can be designed using conventional narrow band radio techniques. A synchronized pseudo noise code generator that drives the receiver's local oscillator frequency synthesizer does de-hopping in the receiver.

2.5 Direct Sequence Spread Spectrum

Most commercial part 15.247 spread spectrum systems transmit an RF signal bandwidth as wide as 20 to 254 times the bandwidth of the information being sent. Various spread spectrum systems have employed RF bandwidths 1000 times their information bandwidth. Common spread spectrum systems are of the 'direct sequence' (Figure 2.1) or 'frequency hopping' type, or else some combination of these two types (called a 'hybrid').

DSSS is an Amplitude Modulation (AM) technique (Figure 2.3). AM has a carrier wave that in this case adjusts signal strength. DSSS spreads the signal by modulating the original signal's waveform with another signal that has a very wide band in relation to the data bandwidth. DSSS prevents high power concentration by spreading the signal over a wide frequency band. The transmitter maps each data bit into a pattern of 'chips'. At the destination, the original data is recreated by mapping the chips back into a bit. The only way that this can operate properly is for the transmitter and receiver to be synchronized. The ratio of chips per bit is called the 'spreading ratio'. A high-spreading ratio increases the resistance of the signal to interference. A low-spreading ratio increases the net bandwidth available to a user. In practice, DSSS spreading ratios are quite small. The majority of 2.4 GHz WLAN product manufacturers offer a spreading ratio of less than 20. A spreading ration of 11 is specified by the IEEE 802.11 standard and the FCC requires that the spreading ratio must be greater than 10.

Several DSSS products in the market allow for the deployment of more than one channel in the same area. They accomplish this by separating the 2.4 GHz band

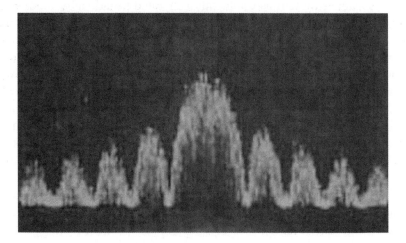

Figure 2.2 A spectrum analyzer photo of a Direct Sequence (DS) spread spectrum signal

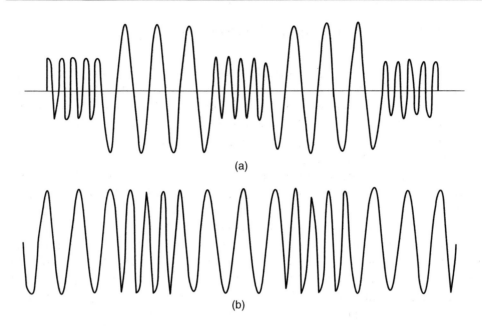

(a)

(b)

Figure 2.3 In amplitude modulation (a), the strength of the signal is modified; in
frequency modulation (b), the frequency of the signal is modified

into several sub-bands, each of which contains an independent DSSS network.
Each DSSS channel occupies 22 MHz of bandwidth. With spectral filtering, three
non-interfering channels spaced 25 MHz apart exist in the 2.4 GHz range. In North
America, these channels are 1, 6, and 11.

A direct sequence system uses a locally generated pseudo noise code to encode
digital data to be transmitted. The local code runs at a much higher rate than the
data rate. Data for transmission is simply logically modulo-2 added (an EXOR
operation) with the faster pseudo noise code. The composite pseudo noise and
data can be passed through a data scrambler to randomize the output spectrum
(and thereby remove discrete spectral lines). A direct sequence modulator is then
used to double sideband suppressed carrier modulate the carrier frequency to be
transmitted. The resultant DSB suppressed carrier AM modulation can also be
thought of as binary phase shift keying (BPSK). Carrier modulation other than
BPSK is possible with direct sequence. However, binary phase shift keying is the
simplest and most often used SS modulation technique.

An SS receiver uses a locally generated replica pseudo noise code and a receiver
correlator to separate only the desired coded information from all possible signals.
An SS correlator can be thought of as a very special matched filter – it responds

only to signals that are encoded with a pseudo noise code that matches its own code. Thus, an SS correlator can be 'tuned' to different codes simply by changing its local code. This correlator does not respond to manmade, natural or artificial noise or interference. It responds only to SS signals with identical matched signal characteristics and encoded with the identical pseudo noise code.

3

Multiple Access Wireless Communications

3.1 Introduction

In multiple access communications systems, which refers to cellular and PCS, there are several goals. These are:

- Universal coverage.
- Quality of Service (QoS) similar to that of traditional telephone service.
- Lower equipment cost for providers and end users.
- Lower number of cell sites required to achieve maximum penetration while maintaining QoS.

The original cellular technology developed in the 1970s by Bell Labs used multiple cells. This original system was called the Advanced Mobile Phone System, or AMPS. It is still in use throughout North America. Similar systems, with slight variations, are Nordic Mobile Telephone (NMT) in Scandinavia, and Total Access Communications System (TACS) used in the United Kingdom, China, and other countries.

The ideal multiple access mobile radio system consists of a family of base stations, or 'cells,' geographically distributed over the service area, and mobile stations.

The spectrum for mobile wireless technology is normally allocated in frequency division duplex (FDD) paired bands. Cellular systems are separated by 45 MHz, while 80 MHz separates PCS bands. There have been some proposals for the use

Wireless Data Technologies. Vern A. Dubendorf
© 2003 John Wiley & Sons, Ltd   ISBN: 0-470-84949-5

of time division duplex (TDD) but such operations inherently limit the coverage area, and have not achieved widespread acceptance.

Traditionally, radio communication systems have separated users by using frequency channels, time slots, or both. These concepts date back to the earliest days of radio. Spark transmitters even used resonant circuits to narrow the spectrum of their radiation. Modern cellular systems began with the use of channelized analog FM.

The 'reuse concept' is familiar even in television broadcasting, where channels are not reused in adjacent cities. This concept of frequency reuse is central to cellular technology. Even though there are hundreds of channels available, if each frequency were assigned to only one cell then the total system capacity would be equal to the total number of channels. When this is adjusted for the Erlang blocking probability, the system could support only a few thousand subscribers. Erlang is a dimensionless unit of telephone traffic intensity named after the Norwegian telephone engineer who first developed the concept. The Erlang is numerically equal to the calling rate times the average holding time. Through reusing channels in multiple cells the system can grow without geographical limits.

3.2 CDMA Overview

From the very beginning, the promise of extraordinary capacity increase over the narrowband multiple access technologies has been the attraction of CDMA. Early models suggested that capacity improvements of more than 20 times would be possible with CDMA as compared to APMS, NMT or TACS. Historically, the capacity was calculated using simple arguments. In reality, it is much more complicated than models. Actual cell coverage areas are very irregular and are not the perfect hexagons that we see in the textbook models. Offered load is not spatially uniform and changes drastically with the time-of-day.

Years ago, Claude Shannon suggested the use of noise-like carrier waves. This is the key to CDMAs high capacity. Instead of partitioning either spectrum or time into disjoint slots, each individual user is assigned a different instance of the noise carrier. While those waveforms are not rigorously orthogonal, they are pretty close. Practical application of this principle has always used digitally generated pseudo-noise, rather than true thermal noise. The basic benefits are preserved, and the transmitters and receivers are simplified because large portions can be implemented using high-density digital devices.

Probably the major benefit of noise-like carriers is that the system sensitivity to interference is fundamentally altered. Traditional time or frequency slotted systems must be designed with a reuse ratio that can satisfy the worst-case interference scenario. Use of noise-like carriers, with all users occupying the same spectrum, makes the effective noise the sum of all other-user signals. The receiver correlates its input with the desired noise carrier, enhancing the signal-to-noise ratio at the detector. The enhancement overcomes the summed noise enough to provide an adequate SNR at the detector. Because the interference is summed, the system is no longer sensitive to worst-case interference, but rather to average interference.

CDMA is a 'spread spectrum' technology, which means that it spreads the information contained in a particular signal of interest over a much greater bandwidth than the original signal. When implemented in a cellular telephone system, CDMA technology offers numerous benefits to the cellular operators and their subscribers. The following is an overview of the benefits of CDMA.

Seven of the benefits provided by CDMA technology include:

1. capacity increases of eight to 10 times that of an AMPS analog system;
2. improved call quality, with better and more consistent sound as compared to AMPS system;
3. simplified system planning through the use of the same frequency in every sector of every cell;
4. enhanced privacy;
5. improved coverage characteristics, allowing for the possibility of fewer cell sites;
6. increased talk time for portables;
7. bandwidth on demand.

3.3 Introduction to CDMA

Code Division Multiple Access (CDMA) is a radical concept in wireless communications. CDMA has gained widespread international acceptance by cellular radio system operators as an upgrade that will dramatically increase both their system capacity and the service quality.

CDMA is a form of *spread-spectrum*, a family of digital communication techniques that have been used in military applications for many years. The core principle of spread spectrum is the use of noise-like carrier waves and, as the name implies, bandwidths much wider than that required for simple point-to-point communication at the same data rate. Originally there were two motivations:

- either to resist enemy efforts to jam the communications (anti-jam, or AJ);
- or to hide the fact that communication was even taking place, sometimes called low probability of intercept (LPI).

It has a history that goes back to the early days of World War II.

The use of CDMA for civilian mobile radio applications is novel. It was proposed theoretically in the late 1940s, but the practical application in the civilian marketplace did not take place until 40 years later. Commercial applications became possible because of two evolutionary developments. One was the availability of very low cost, high density digital integrated circuits, which reduce the size, weight, and cost of the subscriber stations to an acceptably low level. The other was the realization that optimal multiple access communication requires that all user stations regulate their *transmitter* powers to the lowest that will achieve adequate signal quality.

CDMA changes the nature of the subscriber station from a predominately analog device to a predominately digital device. Old-fashioned radio receivers separate stations or channels by filtering in the frequency domain. CDMA receivers do not eliminate analog processing entirely, but they separate communication channels by means of a pseudo-random modulation that is applied and removed in the digital domain, not on the basis of frequency. Multiple users occupy the same frequency band. This universal frequency reuse is not fortuitous. On the contrary, it is crucial to the very high spectral efficiency that is the hallmark of CDMA. Other discussions in these pages show why this is true.

CDMA is altering the face of cellular and PCS communication by:

- dramatically improving the telephone traffic capacity;
- dramatically improving the voice quality and eliminating the audible effects of multipath fading;
- reducing the incidence of dropped calls due to handoff failures;
- providing reliable transport mechanism for data communications, such as facsimile and Internet traffic;
- reducing the number of sites needed to support any given amount of traffic;
- simplifying site selection;
- reducing deployment and operating costs because fewer cell sites are needed;
- reducing average transmitted power;
- reducing interference to other electronic devices;
- reducing potential health risks.

3.4 Principles of CDMA

The goal of spread spectrum technology is to provide a substantial increase in bandwidth of an information-bearing signal, far beyond that needed for basic communication. The bandwidth increase, while not necessary for communication, can mitigate the harmful effects of interference, either deliberate, like a military jammer, or inadvertent, like co-channel users. The interference mitigation is a well-known property of all spread spectrum systems. However, the cooperative use of these techniques in a commercial, non-military environment, to optimize spectral efficiency was a major conceptual advance.

Spread Spectrum systems generally fall into one of two categories: frequency hopping (FH) or direct sequence (DS). In both cases synchronization of transmitter and receiver is required. Both forms can be regarded as using a pseudo-random carrier, but they create that carrier in different ways.

Frequency Hopping is typically accomplished by rapid switching of fast-settling frequency synthesizers in a pseudo-random pattern. The references can be consulted for further discussions of FH, which is not a part of commercial CDMA.

CDMA uses a form of direct sequence. Direct sequence is, in essence, multiplication of a more conventional communication waveform by a pseudo-noise (PN) ± 1 binary sequence in the transmitter. Spreading takes place prior to any modulation, entirely in the binary domain, and the transmitted signals are carefully band limited.

The noise and interference, being uncorrelated with the PN sequence, become noise-like and increase in bandwidth when they reach the detector. Narrowband filtering that rejects most of the interference power can enhance the signal-to-noise ratio. It is often said, with some poetic license, that the SNR is enhanced by the so-called *processing gain* W/R, where W is the spread bandwidth and R is the data rate. This is a partial truth. A careful analysis is needed to accurately determine the performance. In IS-95A CDMA $W/R = 10 \log(1.2288\,\text{MHz}/9600\,\text{Hz}) = 21\,\text{dB}$ for the 9600 bps rate set. To get this right, you have to bite the bullet, and go do some math! We've tried to present it in as simple a fashion as possible.

3.5 Common Air Interface

There are two CDMA common air interface standards:

- Cellular (824–894 MHz) – TIA/EIA/IS-95A
- PCS (1850–1990 MHz) – ANSI J-STD-008.

They are very similar in their features, with the obvious exceptions of the frequency plan, mobile identities, and related message fields. Our purpose here is only to give a general overview and foster a modest level of understanding. For that reason we generally do not distinguish between the two standards. The official documents should always be consulted for precise information.

It should be kept in mind that, while the standards are quite stable documents, they are subject to an ongoing review process by the standards bodies that are responsible for them. Revisions can be expected. There are also many associated documents such as network interface standards, service option standards (primarily vocoders), and performance specifications.

3.6 Forward CDMA Channel

The Forward CDMA Channel is defined as the communication that takes place from the cell to the mobile unit. This communication carries traffic, a pilot signal, and other overhead information. The pilot signal is an unmodulated DSSS signal. The pilot signal and overhead channels establish the systems timing and station identification. The pilot signal or channel is also used in the mobile-assisted handoff (MAHO) process as a signal strength reference.

3.7 Frequency Plan

The base station transmit frequency is 45 MHz above the mobile station transmit frequency in the cellular service (IS-95A), and 80 MHz above in the PCS service (ANSI J-STD-008). Permissible frequency assignments are on 30 kHz increments in cellular services and 50 kHz in PCS services.

3.8 Transmission Parameters

Currently, the IS-95A forward link supports a 9600 bps rate family in the three data-bearing channel types. In all cases the FEC code rate is 1/2 and the PN rate is 1.2288 MHz. Note that 1.2288 MHz = 128 × 9600 bps.

J-STD-008 supports, in addition to the above rates, a second traffic channel rate family with a maximum rate of 14 400 bps. This is termed Rate Set 2, the original 9600 bps family being Rate Set 1. Rate Set 2 uses an FEC code rate of 3/4, created by puncturing the code used in Rate Set 1.

3.9 Overhead Channels

In the forward link there are three types of overhead channels. These are:

- pilot
- sync
- paging.

The pilot is required in every station.

3.9.1 Pilot Channel

The pilot channel is always code channel zero. It serves as the demodulation reference for both the mobile receivers, and for handoff level measurements, and is required to be present in every station. This pilot channel does not carry any information. It is pure short code and has no additional cover or information content.

The pilot channels' amplitude along with its spatial distribution must be carefully controlled due to their relative amplitudes control handoff boundaries between stations. The PN_I and PN_Q modified linear feedback shift register sequences that comprise the short code have period 2^{15} chips, which is $80/3 = 26.667$ ms at the 1.2288 MHz chip rate.

All stations use the same short code, and have the same pilot waveform. The stations are distinguished from one another by the phase of the pilot. The short period of the short code facilitates rapid pilot searches by the mobiles. The air interfaces stipulate that pilot phases always be assigned to stations in multiples of 64 chips, giving a total of $2^{15-6} = 512$ possible assignments. The nine-bit number that identifies the pilot phase assignment is known as the *Pilot Off-set*.

3.9.2 Sync Channel

The sync channel carries a repeating message that identifies the station, and the absolute phase of the pilot sequence. The data rate will always be 1200 bps. The interleaver period is $80/3 = 26.667$ ms, which is equal to the period of the short code. This makes finding frame boundaries much easier once the mobile has located the pilot.

The timing and system configuration information is sent to the mobile station in the Sync Channel using a repeating message. The mobile station can calculate accurate system time by synchronizing to the short code. The short code synchronization and the pilot offset, which is part of the sync message, fix system time modulo 26.667 ms. The short code period ambiguity is then resolved by the long code state and system time fields that are also part of the sync message.

3.9.3 Paging Channel

The paging channel is used for communications with mobile stations when the mobile station has not been assigned to a traffic channel. The primary purpose of the paging channel is to convey pages or notifications of incoming calls being sent to the mobile stations. It carries the responses to mobile station accesses, both page responses and unsolicited originations. Once there is a successful access, this is normally followed by an assignment to a dedicated traffic channel. After a traffic channel has been established then signaling traffic between base and mobile can continued interspersed with the user traffic.

The paging channel may run at either 4800 or 9600 bps.

Every base station is required to have at least one paging channel per sector and it must be on at least one of the frequencies in use. All paging can be done on one frequency, or it can be distributed over multiple frequencies.

3.9.4 Traffic Channel

Traffic channels are assigned dynamically to mobile stations in response to mobile station accesses. Through the use of a paging channel message, the mobile station is told which code channel it is to receive.

The traffic channel always carries data in 20 ms frames. Frames at the higher rates of Rate Set 1, and in all frames of Rate Set 2, include CRC codes to help assess the frame quality in the receiver.

3.10 Soft Handoff

What happens during soft handoffs is that each base station that is participating in the handoff transmits the same traffic over its assigned code channel. The code channel assignments are independent, and are normally different in each cell. Whatever code channels are not in use for overhead channels are available, up to either a total of 64 or the available equipment limit, whichever is smaller.

3.11 Rate

Traffic channels carry variable rate traffic frames, either 1, 1/2, 1/4, or 1/8 of the maximum rate. In IS-95A only a 9600 bps rate family is currently available in the standard. In J-STD-008 a second rate set, based on a maximum rate of 14 400 bps is available. The Rate Set 2 will be added in a future revision of IS-95.

One, two, four, or eight-way repetition of code symbols accomplishes the rate variation. Transmission is continuous, with the amplitude reduced at the lower rates in order to keep the energy per bit approximately constant, regardless of rate. The rate is independently variable in each 20 ms frame.

3.12 Power Control Subchannel

The 800 bps reverse link power control subchannel is carried on the traffic channel by puncturing two from every 24 symbols transmitted. Both of the punctured symbols carry the same power control bit in order for them to be coherently combined by the receiver. Each base station participating in a soft handoff makes its own power control decision, which is totally independent of the others unless they are different sectors of the same cell. If they are different sectors of the same cell, they all transmit a common decision. This special circumstance is made known to the mobile when the handoff is set up.

3.13 Timing

All base stations must be synchronized within a few microseconds for the station identification mechanisms to work.

3.14 Reverse CDMA Channel

The mobile-to-cell direction of communication is known as the Reverse CDMA Channel. This channel carries traffic and signaling. Reverse channels are only active during calls to the associated mobile station and also when access channel signaling is taking place to the associated base station.

3.14.1 Frequency Plan

The mobile station transmit frequency is 45 MHz below the base station transmit frequency in the cellular service (IS-95A), and 80 MHz below in the PCS service (ANSI J-STD-008). Permissible frequency assignments are on 30 kHz increments in cellular and 50 kHz in PCS.

3.14.2 Transmission Parameters

The IS-95A Reverse CDMA Channel currently supports a 9600 bps rate family in the Access Channel and Traffic Channels. The transmission duty cycle varies with data rate. In all cases the FEC code rate is 1/3, the code symbol rate is always 28 800 symbols per second after there are six code symbols per modulation symbol, and the PN rate is 1.2288 MHz. The modulation is 64-ary orthogonal which uses the same Walsh functions used in the forward link for channelization.

3.15 Signal Structure

3.15.1 Channelization

The Reverse CDMA Channel consists of $2^{42} - 1$ logical channels with one of these logical channels being permanently associated with each mobile station. The mobile station then uses these logical channels when it passes traffic. The channel does not change upon handoff. Other logical channels are also associated with base stations for system access.

3.15.2 Separation of Users

Reverse CDMA Channels do not use strict orthogonality to separate logical channels, as do the Forward CDMA Channels. Instead it uses very long period spreading codes in distinct phases.

3.15.3 Orthogonal Modulation

Reverse link data modulation is 64-ary orthogonal, which is applied prior to the spreading. Groups of six code symbols select one of 64 orthogonal sequences. The 64-ary orthogonal sequences are the same Walsh functions that are used in the Forward CDMA Channel, but are used here for a totally different purpose.

3.15.4 Traffic Channel

Traffic channels are located in the Reverse CDMA Channel. The traffic channel always carries data in 20 ms frames. The majority of the frames, with a few exceptions, also include CRC codes to help assess the frame quality in the receiver.

3.15.5 Soft Handoff

There is no change in the content of the mobile transmissions during a soft handoff. There is however a change in the way that reverse link power control is applied to the reverse link.

Power control 'down' commands from all participants in the handoff are logically 'or'ed' together – if any of the handoff participant base stations says 'down,' then the mobile station is required to reduce its reverse link power.

3.15.6 Rate

Traffic channels carry variable rate traffic frames, either 1, 1/2, 1/4, or 1/8 of the maximum rate. In IS-95A only a 9600 bps rate family is currently available in the standard. In J-STD-008 a second rate set, based on a maximum rate of 14 400 bps is available.

The data rate variation is accomplished by varying the duty cycle of transmission in accordance with a 1, 1/2, 1/4, 1/8 plan, according to the rate requested by the data source. Transmission always occurs in 1.25 ms segments. The segments that are actually transmitted are pseudo-randomly selected, using a decimated long code sequence.

3.15.7 *Timing*

All mobile stations are required to adjust their transmission time according to the timing that they are able to derive from the pilot and sync channels, and adjusted for the know base station pilot offset.

3.16 TDMA

Time Division Multiple Access or TDMA is defined as a technology for delivering digital wireless service using time-division multiplexing (TDM). TDMA works by dividing a radio frequency into time slots and then allocating slots to multiple calls. In this way, a single frequency can support multiple, simultaneous data channels. TDMA is used by the GSM digital cellular system. The TDMA digital transmission scheme multiplexes three signals over a single channel. The current TDMA standard for cellular divides a single channel into six time slots, with each signal using two slots, providing a three-to-one gain in capacity over advanced mobile-phone service (AMPS). Each caller is assigned a specific time slot for transmission.

In the mid-to-late 1980s the wireless industry began to explore converting the existing analog network to digital in order to improve capacity. In 1989, the Cellular Telecommunications Industry Association (CTIA) chose TDMA over Motorola's frequency division multiple access (FDMA) narrowband standard as the technology of choice for existing 800 MHz cellular markets and for emerging 1.9 GHz markets.

With the growing technology competition applied by Qualcomm in favor of code division multiple access (CDMA) and the realities of the European global system for mobile communications (GSM) standard, the CTIA decided to let carriers make their own technology selection.

The TDMA system is designed for use in a range of environments and situations, from hand portable use in a downtown office to a mobile user traveling at high speed on the freeway. TDMA systems also provide support for a variety of services for the end user, such as voice, data, fax, short message services, and broadcast messages. TDMA provides a flexible air interface that provides high performance with respect to capacity, coverage, and unlimited support of mobility and capability to handle different types of user needs.

All of the multiple access techniques now depend on the adoption of digital technology. Digital technology has become the standard for the public telephone system. In this system, also known as PSTN, all analog calls are converted to digital form for transmission over the backbone. Digital technology has a number of advantages over analog transmission.

- It economizes on bandwidth.
- It allows easy integration with personal communication systems (PCS) devices.
- It maintains superior quality of voice transmission over long distances.
- It is difficult to decode.
- It can use lower average transmitter power.
- It enables smaller and less expensive individual receivers and transmitters.
- It offers voice privacy.

TDMA requires that the audio signals have already been digitized. It then allocates a single frequency channel for a short time and then moves to another channel. The digital samples from a single transmitter occupy different time slots in several bands at the same time. The access technique used in TDMA has three users sharing a 30 kHz carrier frequency. A single channel can carry multiple conversations if each conversation is divided into relatively short fragments, is assigned a time slot, and is transmitted in synchronized timed bursts.

The IS-54 and IS-136 implementations of TDMA immediately tripled the capacity of cellular frequencies by dividing a 30 kHz channel into three time slots thus enabling three different users to occupy it at the same time.

In addition to increasing the efficiency of transmission, TDMA offers other advantages over standard cellular technologies. It can be easily adapted to the transmission of data as well as voice communication. TDMA offers the ability to carry data rates of 64 kbps to 120 Mbps (expandable in multiples of 64 kbps). This enables operators to offer personal communication-like services including fax, voiceband data, and short message services (SMSs) as well as bandwidth-intensive applications such as multimedia and video-conferencing.

Spread spectrum techniques can suffer from interference among the users all of whom are on the same frequency band and transmitting at the same time. TDMA's technology, which separates users in time, ensures that they will not experience interference from other simultaneous transmissions and does not have the same problems as spread spectrum. Another benefit of TDMA is that it also provides the user with extended battery life and talk time since the mobile is only transmitting a portion of the time (from 1/3 to 1/10) of the time during conversations.

3.16.1 TDMA Standards

The original TDMA standard was IS-54 which was introduced in 1988–1989 by the Telecommunications Industry Association (TIA)/CTIA. It provided a feature set, which included authentication, calling-number ID, a message-waiting indicator (MWI), and voice privacy.

- IS-54B was superseded in 1994 with the introduction of IS-136 followed closely by revisions A and B.
- IS-136 was backward compatible to IS-54B and included a DCCH and advanced features.
- IS-136A upbanded IS-136 for seamless cellular service between the 800 MHz and 1900 MHz frequency bands. In addition, it introduced over-the-air activation and programming services.
- IS-136B includes a new range of services including broadcast SMS, packet data and others.

4

GSM

4.1 Introduction

It may be hard to realize this, but it really was not all that long ago when just a plain old telephone was a luxury item. Nevertheless, as we all know, technology's only constant is change. Currently, many folks need to be accessible everywhere, whether they are at work or play, in the office or at home. To meet this demand, the GSM standard (Global System for Mobile Communications) for mobile telephony was introduced in the mid-1980s.

Today, GSM is the most popular mobile radio standard in the world. A boom is underway, and this is demonstrated by the fact that many GSM users find life without their phone practically inconceivable.

Nowadays, when we speak of GSM, we usually mean 'original' GSM also known as GSM900 since 900 MHz was the original frequency band. To provide additional capacity and enable higher subscriber densities, two other systems were added later: GSM1800 (also DCS1800) and GSM1900 (also PCS 900). Compared to GSM 900, GSM1800 and GSM1900 differ primarily in the air interface. Besides using another frequency band, they use a microcellular structure (i.e. a smaller coverage region for each radio cell). This makes it possible to reuse frequencies at closer distances, enabling an increase in subscriber density. The disadvantage is the higher attenuation of the air interface due to the higher frequency.

Where now? A few years ago, Michael Jackson sang '. . . just call my name and I'll be there'. While this might seem inconceivable now, it might become reality sooner than we think, given the rapid pace of technological evolution. Faced with a whirlwind of speculation, ETSI (the telecom standardization authority in

Wireless Data Technologies.   Vern A. Dubendorf
© 2003 John Wiley & Sons, Ltd   ISBN: 0-470-84949-5

Europe) decided to base the air interface of the planned universal mobile telecommunications system (UMTS) on a mix of WCDMA and TD/CDMA technologies. The infrastructure of the existing GSM networks will most likely be used.

This chapter is intended to provide basic information about the GSM system.

4.2 Overview

Before GSM networks there were public mobile radio networks (cellular). They normally used analog technologies, which varied from country to country and from one manufacturer to another. These analog networks did not comply with any uniform standard. There was no way to use a single mobile phone from one country to another. The speech quality in most networks was not satisfactory. For an overview of the mobile evolution see Figure 4.1 above.

GSM became popular very quickly because it provided improved speech quality and, through a uniform international standard, made it possible to use a single telephone number and mobile unit around the world. The European Telecommunications Standardization Institute (ETSI) adopted the GSM standard in 1991, and GSM is now used in 135 countries.

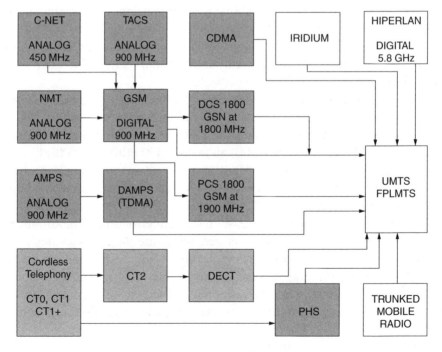

Figure 4.1   The mobile evolution

The benefits of GSM include:

- support for international roaming;
- distinction between user and device identification;
- excellent speech quality;
- wide range of services;
- interworking (e.g. with ISDN, DECT);
- extensive security features.

GSM also stands out from other technologies with its wide range of services; available services vary from operator to operator:

- telephony;
- asynchronous and synchronous data services (2.4/4.8/9.6 kbit/s);
- access to packet data network (X.25);
- telematic services (SMS, fax, videotext, etc.);
- many value-added features (call forwarding, caller ID, voice mailbox);
- e-mail and Internet connections.

The best way to create a manageable communications system is to divide it into various subgroups that are interconnected using standardized interfaces. A GSM network can be divided into three groups (see Fig. 4.2): The mobile station (MS), the base station subsystem (BSS) and the network subsystem.

They are characterized as follows:

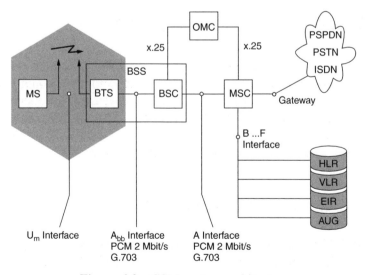

Figure 4.2   GSM system architecture

4.2.1 The Mobile Station (MS)

A mobile station may be referred to as a 'handset', a 'mobile', a 'portable terminal' or 'mobile equipment' (ME).

It also includes a subscriber identity module (SIM) that is normally removable and comes in two sizes. Each SIM card has a unique identification number called international mobile subscriber identity (IMSI).

In addition, each MS is assigned a unique hardware identification called international mobile equipment identity (IMEI). In some of the newer applications (data communications in particular), an MS can also be a terminal that acts as a GSM interface, e.g. for a laptop computer. In this new application, the MS does not look like a normal GSM telephone.

The seemingly low price of a mobile phone can give the (false) impression that the product is not of high quality. Besides providing a transceiver (TRX) for transmission and reception of voice and data, the mobile also performs a number of very demanding tasks such as authentication, handover, encoding and channel encoding.

4.2.2 The Base Station Subsystem (BSS)

The base station subsystem (BSS) is made up of the base station controller (BSC) and the base transceiver station (BTS).

4.2.3 The Base Transceiver Station (BTS)

GSM uses a series of radio transmitters called BTSs to connect the mobiles to a cellular network. Their tasks include channel coding/decoding and encryption/decryption. A BTS is comprised of radio transmitters and receivers, antennas, the interface to the PCM facility, etc. The BTS may contain one or more transceivers to provide the required call handling capacity. A cell site may be omnidirectional or split into typically three directional cells.

4.2.3.1 The Base Station Controller (BSC)

A group of BTSs are connected to a particular BSC that manages the radio resources for them. Today's new and intelligent BTSs have taken over many tasks that were previously handled by the BSCs.

The primary function of the BSC is call maintenance. The mobile stations normally send a report of their received signal strength to the BSC every 480 ms. With this information the BSC decides to initiate handovers to other cells, change the BTS transmitter power, etc.

4.2.4 The Network Subsystem

The network subsystem is made up of the following five units.

4.2.4.1 The Mobile Switching Center (MSC)

The mobile switching center (MSC) acts as a standard exchange in a fixed network and additionally provides all the functionality needed to handle a mobile subscriber. The main functions are registration, authentication, location updating, handovers and call routing to a roaming subscriber. The signaling between functional entities (registers) in the network subsystem uses Signaling System 7 (SS7). If the MSC also has a gateway function for communicating with other networks, it is called Gateway MSC (GMSC).

4.2.4.2 The Home Location Register (HLR)

A database used for management of mobile subscribers. It stores the international mobile subscriber identity (IMSI), mobile station ISDN number (MSISDN) and current visitor location register (VLR) address. The main information stored there concerns the location of each mobile station in order to be able to route calls to the mobile subscribers managed by each HLR. The HLR also maintains the services associated with each MS. One HLR can serve several MSCs.

4.2.4.3 The Visitor Location Register (VLR)

Contains the current location of the MS and selected administrative information from the HLR, necessary for call control and provision of the subscribed services, for each mobile currently located in the geographical area controlled by the VLR. A VLR is connected to one MSC and is normally integrated into the MSC's hardware.

4.2.4.4 The Authentication Center (AuC)

A protected database that holds a copy of the secret key stored in each subscriber's SIM card, which is used for authentication and encryption over the radio channel. The AuC provides additional security against fraud. It is normally located close to each HLR within a GSM network.

4.2.4.5 The Equipment Identity Register (EIR)

The EIR is a database that contains a list of all valid mobile station equipment within the network, where each mobile station is identified by its international mobile equipment identity (IMEI). The EIR has three databases:

- white list – for all known, good IMEIs;
- black list – for bad or stolen handsets;
- gray list – for handsets/IMEIs that are uncertain.

4.2.5 The Operation and Maintenance Center (OMC)

The OMC is a management system that oversees the GSM functional blocks. The OMC assists the network operator in maintaining satisfactory operation of the GSM network. Hardware redundancy and intelligent error detection mechanisms help prevent network downtime. The OMC is responsible for controlling and maintaining the MSC, BSC, and BTS. It can be in charge of an entire public land mobile network (PLMN) or just some parts of the PLMN.

4.3 Interfaces and Protocols

Providing voice or data transmission quality over the radio link is only part of the function of a cellular mobile network. A GSM mobile can seamlessly roam nationally and internationally, requiring standardized call routing and location updating functions in GSM networks. A public communications system also needs solid security mechanisms to prevent misuse by third parties. Security functions such as authentication, encryption, and the use of Temporary Mobile Subscriber Identities (TMSIs) are an absolute must.

4.3.1 Protocols

Within a GSM network, different protocols are needed to enable the flow of data
and signaling between different GSM subsystems. Figure 4.3 shows the interfaces
that link the different GSM subsystems and the protocols used to communicate
on each interface.

GSM protocols are basically divided into three layers:

- Layer 1 – Physical Layer
  - Enables physical transmission (TDMA, FDMA, etc.)
  - Assessment of channel quality
  - Except of the air interface (GSM Rec. 04.04), PCM 30 or ISDN links are
    used (GSM Rec. 08.54 on A$_{bis}$ interface and 08.04 on A to F interfaces).
- Layer 2 – Data Link Layer
  - Multiplexing of one or more layer-two connections on control/signaling
    channels
  - Error detection (based on HDLC)
  - Flow control
  - Transmission quality assurance
  - Routing.
- Layer 3 – Network Layer
  - Connection management (air interface)
  - Management of location data

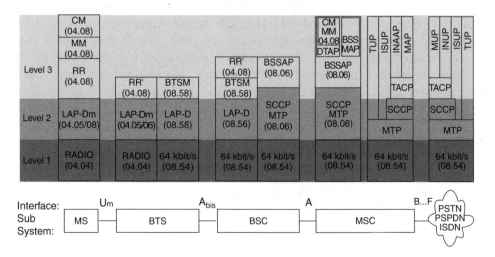

Figure 4.3   OSI layer structure in GSM

- Subscriber identification
- Management of added services (SMS, call forwarding, conference calls, etc.)

4.3.2 The Air Interface

The International Telecommunication Union (ITU), which manages international allocation of radio spectrum (among many other functions), has allocated the following bands.

- GSM900:
  - Uplink: 890–915 MHz (= mobile station to base station)
  - Downlink: 935–960 MHz (= base station to mobile station).
- GSM1800 (previously: DCS-1800):
  - Uplink: 1710–1785 MHz
  - Downlink: 1805–1880 MHz.
- GSM1900 (previously: PCS-1900):
  - Uplink: 1850–1910 MHz
  - Downlink: 1930–1990 MHz.

The air interface for GSM is known as the U_m interface.

Since radio spectrum is a limited resource shared by all users, a method was devised to divide the bandwidth among as many users as possible. The method chosen by GSM is a combination of time and frequency-division multiple access (TDMA/FDMA). The FDMA part involves the division by frequency of the (maximum) 25 MHz allocated bandwidth into 124 carrier frequencies spaced 200 kHz apart. One or more carrier frequencies are assigned to each base station. Each of these carrier frequencies is then divided in time, using a TDMA scheme. The fundamental unit of time in this TDMA scheme is called a burst period and it lasts approximately 0.577 ms. Eight burst periods are grouped into a TDMA frame (approximately 4.615 ms), which forms the basic unit for the definition of logical channels. One physical channel is one burst period per TDMA frame. (Fig. 4.4 GSM air interface, TDMA frame).

4.3.2.1 Protocols on the Air Interface

Layer 1 (GSM Rec. 04.04)

The physical properties of the U_m interface have already been described.

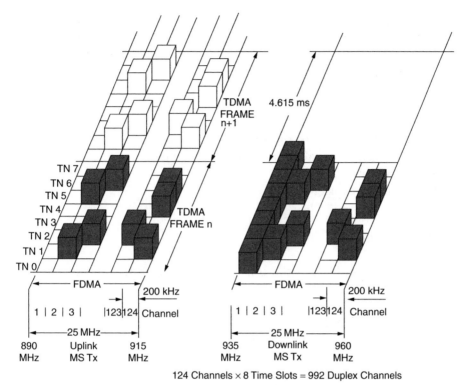

Figure 4.4   GSM air interface, TDMA frame

Layer 2 (GSM Rec. 04.05/06)

Here, the LAP-Dm protocol is used (similar to ISDN LAP-D). LAP-Dm has the following functions:

- connectionless transfer on point-to-point and point-to-multipoint signaling channels;
- setup and take-down of layer-two connections on point-to-point signaling channels;
- connection-oriented transfer with retention of the transmission sequence, error detection, and error correction.

Layer 3 (GSM Rec. 04.07/08)

Contains the following sublayers which control signaling channel functions (BCH, CCCH and DCCH):

Radio resource management (RR). The role of the RR management layer is to establish and release stable connection between mobile stations (MS) and an MSC for the duration of a call, and to maintain it despite user movements.

The following functions are performed by the MSC:

- cell selection;
- handover;
- allocation and take-down of point-to-point channels;
- monitoring and forwarding of radio connections;
- introduction of encryption;
- change in transmission mode.

Mobility management (MM)

Mobility management handles the control functions required for mobility, e.g.:

- authentication;
- assignment of TMSI;
- management of subscriber location.

Connection management (CM)

Connection management is used to set up, maintain, and take down calls connections; it is comprised of the following three subgroups.

- call control (CC): manages call connections;
- supplementary service support (SS): handles special services;
- short message service support (SMS): transfers brief texts.

Neither the BTS nor the BSC interpret CM and MM messages. They are simply exchanged with the MSC or the MS using the direct transfer application part (DTAP) protocol on the A interface. RR messages are mapped to or from the base station system application part (BSSAP) in the BSCREF for exchange with the MSC.

4.3.3 Logical Channels on the Air Interface

Several logical channels are mapped onto the physical channels. (Fig. 4.5 GSM air interface, logical channels). The organization of logical channels depends on

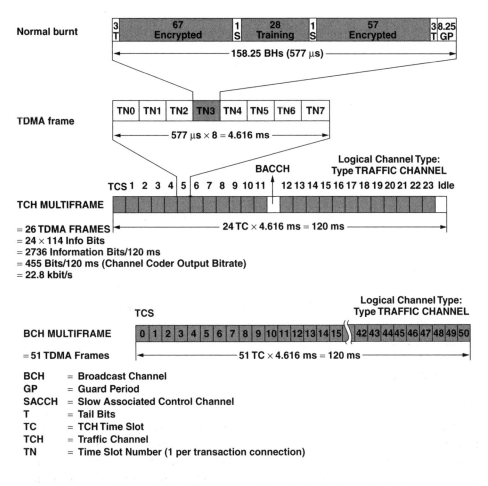

Figure 4.5 GSM air interface, logical channels

the application and the direction of information flow (uplink/downlink or bi-directional). A logical channel can be either a traffic channel (TCH), which carries user data, or a signaling channel (see following chapters).

4.3.4 Traffic Channels on the Air Interface

A traffic channel (TCH) is used to carry speech and data traffic. Traffic channels are defined using a 26-frame multiframe, or group of 26 TDMA frames. The length of a 26-frame multiframe is 120 ms, which is how the length of a burst period is defined (120 ms divided by 26 frames divided by eight burst periods per frame). Out of the 26 frames, 24 are used for traffic, one is used for

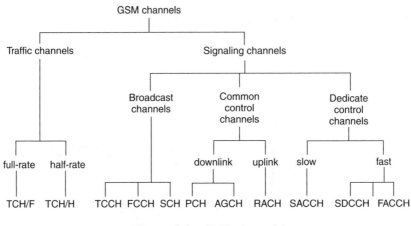

Figure 4.6   GMS channels

the slow associated control channel (SACCH) and one is currently unused (see Figure 4.6 GMS channels). TCHs for the uplink and downlink are separated in time by three burst periods, so that the mobile station does not have to transmit and receive simultaneously, thereby simplifying the electronic circuitry. This method permits complex antenna duplex filters to be avoided and thus helps to cut power consumption.

In addition to these full-rate TCHs (TCH/F, 22.8 kbit/s), half-rate TCHs (TCH/H, 11.4 kbit/s) are also defined. Half-rate TCHs double the capacity of a system effectively by making it possible to transmit two calls in a single channel. If a TCH/F is used for data communications, the usable data rate drops to 9.6 kbit/s (in TCH/H: max. 4.8 kbit/s) due to the enhanced security algorithms. Eighth rate TCHs are also specified, and are used for signaling. In the GSM Recommendations, they are called stand-alone dedicated control channels (SDCCH).

4.3.5 Signaling Channels on the Air Interface

The signaling channels on the air interface are used for call establishment, paging, call maintenance, synchronization, etc. There are three groups of signaling channels as follows.

4.3.5.1 The Broadcast Channels (BCH)

The broadcast channels carry only downlink information and are responsible mainly for synchronization and frequency correction. This is the only channel type enabling point-to-multipoint communications in which short messages are simultaneously transmitted to several mobiles.

The BCHs include the following channels:

- The broadcast control channel (BCCH): general information, cell-specific; e.g. local area code (LAC), network operator, access parameters, list of neighboring cells, etc. The MS receives signals via the BCCH from many BTSs within the same network and/or different networks.
- The frequency correction channel (FCCH): downlink only; correction of MS frequencies; transmission of frequency standard to MS; it is also used for synchronization of an acquisition by providing the boundaries between timeslots and the position of the first time slot of a TDMA frame.
- The synchronization channel (SCH): downlink only; frame synchronization (TDMA frame number) and identification of base station. The valid reception of one SCH burst will provide the MS with all the information needed to synchronize with a BTS.

4.3.5.2 The Common Control Channels (CCCH)

The common control channels are a group of uplink and downlink channels between the MS card and the BTS. These channels are used to convey information from the network to MSs and provide access to the network.

The CCCHs include the following channels:

- The paging channel (PCH): downlink only; the MS is informed by the BTS for incoming calls via the PCH.
- The access grant channel (AGCH): downlink only; BTS allocates a TCH or SDCCH to the MS, thus allowing the MS access to the network.
- The random access channel (RACH): uplink only; allows the MS to request an SDCCH in response to a page or due to a call; the MS chooses a random time to send on this channel. This creates a possibility of collisions with transmissions from other MSs.

The PCH and AGCH are transmitted in one channel called the paging and access grant channel (PAGCH). They are separated by time.

4.3.5.3 The Dedicated Control Channels (DCCH)

The dedicated control channels are responsible for roaming, handovers, encryption, etc.

The DCCHs include the following channels:

- The stand-alone dedicated control channel (SDCCH): communications channel between MS and the BTS; signaling during call setup before a traffic channel (TCH) is allocated.
- The slow associated control channel (SACCH): transmits continuous measurement reports (e.g. field strengths) in parallel to operation of a TCH or SDCCH; needed, e.g. for handover decisions; always allocated to a TCH or SDCCH; needed for 'non-urgent' procedures, e.g. for radio measurement data, power control (downlink only), timing advance, etc.; always used in parallel to a TCH or SDCCH.
- The fast associated control channel (FACCH): similar to the SDCCH, but used in parallel to operation of the TCH; if the data rate of the SACCH is insufficient, 'borrowing mode' is used. 'Borrowing mode' is where additional bandwidth is borrowed from the TCH; this happens for messages associated with call establishment authentication of the subscriber, handover decisions, etc.

Almost all of the signaling channels use the 'normal burst' format (see the GSM section on Burst formats), except for the RACH (Random Access Burst), FCCH (Frequency Correction Burst) and SCH (SynCHronization Burst) channels.

4.3.6 Burst Formats

A timeslot is a 576 ms time interval, i.e. 156.25 bits duration, and its physical contents are known as a burst. Five different types of bursts exist in the system. Different TDMA frame divisions distinguish them.

The normal burst (NB)

Used to carry information on traffic and control channels, except for RACH. It contains 116 encrypted bits.

The frequency correction burst (FB)

Used for frequency synchronization of the mobile. The contents of this burst are used to calculate an unmodulated, sinusoidal oscillation, onto which the synthesizer of the mobiles is clocked.

The synchronization burst (SB)

Used for time synchronization of the mobile. It contains a long training sequence and carries the information of a TDMA frame number.

The access burst (AB)

Used for random access and characterized by a longer guard period (256 ms) to allow for burst transmission from a mobile that does not know the correct timing advance at the first access to a network (or after handover).

The dummy burst (DB)

Transmitted as a filler in unused timeslots of the carrier; does not carry any information but has the same format as a normal burst (NB).

5

GPRS (General Packet Radio Service) for GSM

5.1 Introduction

Wireless data usage has been doubling every year for the past several years in advanced market regions. Several of the cellular operators currently earn in excess of 6% of their revenues from the wireless data market. GPRS is, overall, the best platform for mobile data networking services. An essential stepping stone, GPRS is the lead-in to 3G or third-generation multimedia services. While opening new opportunities for the operators, GPRS also introduces new challenges; the most significant of these are changes to the tariff model and introduction of the IP (Internet Protocol) infrastructure.

GPRS is the packet-mode extension to GSM. It uses a new physical channel called a 52-multiframe that is made up of two 26-control multiframes of voice-mode GSM but continues to use the same air interface. The packet-mode controls as well as the data channels are mapped into different slots of the 52-multiframe.

The GPRS architecture is shown in Figure 5.1. The *SGSN* or servicing GPRS support node is in charge of one or more GPRS base stations (BSs). The base station controller (BSC) monitors and controls several BSs, or base transceiver stations (BTSs). The BSC and its BTSs form a base station subsystem (BSS). A BTS and the MSs in its control form a cell. The gateway GPRS support node (GGSN) is for interconnection to the Internet. The registration of packet mode MSs is done with the architectural entities, the MSC/VLR (mobile switching center and visiting location register) and home locator register (HLR).

Logical channels of GPRS are the packet common control channels (PCCCHs). They are comprised of logical channels for common control signaling used for

Wireless Data Technologies. Vern A. Dubendorf
© 2003 John Wiley & Sons, Ltd ISBN: 0-470-84949-5

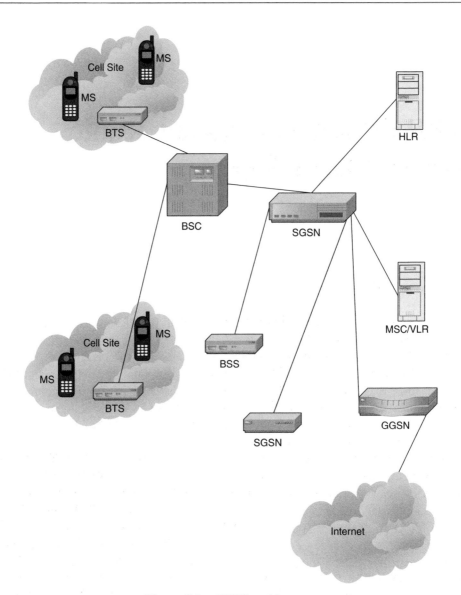

Figure 5.1   GPRS architecture

packet data and a packet random access channel (PRACH) uplink only, which is used by an MS to initiate uplink transfer for sending data or signaling information, a packet paging channel (PPCH) downlink only, which is used to page an MS prior to downlink packet transfer, and a packet access grant channel (PAGCH) downlink only, which is used in the packet transfer establishment phase to send resource assignment to an MS prior to packet transfer.

There are three key features that describe GPRS:

(1) always online;
(2) an upgrade to current systems;
(3) an integral part of future 3rd Generation (3G) Systems.

In a cost-effective move, wireless operators are launching packet data services over mobile networks around the globe. Deploying packet data provides a method for mobile carriers to balance the network resources needed to sufficiently meet the needs of the growing market for voice services and the demand generated by the expanding mobile data market.

GPRS will require Mobile IP in order to offer full mobility within the Internet. Without Mobile IP capabilities the GPRS network will not be able to identify a node such as a portable computer that has a standard IP address. Mobile IP is required to allow the corporate network to authenticate the IP address of the portable computer.

5.2 Always Online

GPRS allows the user to be constantly connected, always online, without having to pay by the minute, very similar in which we use the features of DSL (Digital Subscriber Line).

5.3 Differences between GPRS/GSM and cdmaOne

5.3.1 GSM

A new data backbone is required by GSM operators. Base station upgrades and new handsets to offer packet data services are also required. GSM is circuit-based architecture that requires a new packet data backbone and new handsets. Each of the future GSM data services, which include EDGE and WCDMA, will require the purchase of new mobile phones in order to take advantage of the enhanced functionality. We also know that the GSM roadmap for handsets is not forward or backward compatible. What this means is that GPRS handsets will not work on EDGE or 3G CDMA DS base stations.

So we know that a GSM carrier will be required to implement and IP-based network that will allow a packet overlay onto circuit switched networks. The links between the existing GSM network infrastructure and the IP backbone are comprised of proprietary hardware such as the Gateway GPRS Service Nodes (GGSNs) which link the Internet to the IP backbone. GGSNs are modified IP routers. One of the issues with GGSNs is that new pieces of equipment raise security concerns with IT departments. Discussions with suppliers of both standard IP routers and GGSNs have indicated that a GGSN will normally cost three to four times more than the equivalent IP router. This can hinder the deployment of a mobile data application due to the need for integration and testing. They will also have to make new investments in base stations for GPRS, EDGE and 3G CDMA implementations, however the IP-based backbone will require only minor modifications after initially deploying the GPRS network.

5.3.2 cdmaOne IS-95

cdmaOne is based on IP standards, giving it an inherent advantage over GPRS. Today, all cdmaOne handsets and base stations are packet data capable, and the networks utilize Internet protocol (IP)-based equipment. This allows for a high degree of backward and forward hardware compatibility for network operators looking to implement new higher speed data services and evolve to 3G, which is an IP-based standard. These cdmaOne networks already incorporate an IP gateway referred to as the Inter-Working Function (IWF). This is essentially a standard IP router built into the network, routing IP packets without the need for them to be handled by an analog modem. The IWF receives information from the mobile phone in Point-to-Point Protocol (PPP) format and assigns a temporary IP address for that session. Network infrastructure manufacturers state that their cdmaOne infrastructure allows the incorporation of any standard router from any manufacturer into the IWF. A standard RADIUS server provides billing information and authentication services in the network while messaging is handled using SMTP.

The cdmaOne packet data implementation utilizes standard routers, which are the same ones used in the landline Internet. IT professionals working on a corporate landline Intranet have the ability to transfer the same skill set to a mobile Intranet based on cdmaOne.

5.3.3 Analysis

We can see that the packet data design which is standardized in the network and handsets of cdmaOne technology facilitate easier, less expensive packet data implementation than GPRS from a network operator, handset, application developer and corporation's perspective.

6

iMode

'"i" stands for information, interactive, Internet, and 'ai', which means love in Japanese.'

Keiji Tatekawa, a CEO of NTT DoCoMo

6.1 Introduction

iMode is NTT DoCoMo's mobile Internet access system. 'iMode' is also a trademark and/or service mark owned by NTT DoCoMo.

iMode proves customers with quick continuous connection to the Internet. The success of iMode is now changing the information systems of existing businesses. Customers want direct access to their sales representatives and customized information they can read with their smart phones. Management must change the information system to meet these needs with four standards: accessibility, personalization, information sharing, and authority decentralization.

6.2 What is iMode?

iMode is a brand and a service of wireless Internet connection at the touch of a button on a cell phone. NTT DoCoMo started this service in February 1999 in Japan and obtained over six million subscribers 14 month later.

Users receive news and stock prices tailored their needs, send and receive e-mail, shop online, and receive advice on good restaurants in unfamiliar parts of town. One hundred and sixty financial institutions joined online banking through

Wireless Data Technologies. Vern A. Dubendorf
© 2003 John Wiley & Sons, Ltd   ISBN: 0-470-84949-5

iMode and the latest interface with car navigation system provides congestion news, localized weather forecasts and parking updates for the ultimate in traffic information.

iMode's critical success factors are low costs and quick continuous access to the Internet. iMode charges for the amount of data not for additional minutes. A user pays a fixed charge of $3 per month and 30 cents per packet. Thus, the Internet connection can remain free of charge until the user sends or receives data. European users can tap onto the Internet from their cell phones and Americans get similar benefits with Palm VII devices. However, these systems must set up a new dial-up connection each time a user wants to go onto the Internet.

6.2.1 What does a Typical iMode Screen look like?

The following image shows the screen of Eurotechnology Japan K.K. company's iMode site on a Sony SO503i handset (TFT screen, 65 536 colors). There are in the neighborhood of 70 or more different styles of DoCoMo handsets that includes counting color variations and many more for competing mobile Internet systems in Japan.

6.3 Technology

iMode consists of three technologies: a smart handset, a new transmission protocol, and a new markup language.

6.3.1 Smart Phone

A current high-end cell phone is now equivalent to a low-end PC. It has a 100 MHz processor, many megabytes of flash memory, and a color display with a graphical user interface. These 'smart' phones enable users to brows the Internet with the touch of a button. But users cannot talk while browsing the web. They switch to the web by hitting a URL button on the phone. There is no *defacto* standard in the operation system and browsing software, such as Windows 2000 or Internet Explorer. Since the information that iMode deals with is still simple, each cell phone maker adopts its own system.

6.3.2 Transmission System

6.3.2.1 CDMA

The transmission protocol of iMode is Code Division Multiple Access (CDMA), which enables several subscribers to use the same line at once. iMode's transmission speed is 9.6 Kbps (bit per second), which is slower than a typical modem for personal use, 28.8 Kbps. Thus, email is limited to about 250 characters per message. Although 9.6 Kbps is insufficient to download video, it is appropriate for short e-mail and simple graphics.

6.3.3 Markup Language

6.3.3.1 C-HTML and WAP

iMode adopted C-HTML (compact HTML) as its markup language. Compact HTML is a subtext of HTML, which focuses on text and simple graphics. Since a cell phone has a small display with touch button manipulation, it requires a special markup language to display data. We have two major ways to meet this need: compact HTML and WAP (Wireless Application Protocol). Web site operators can easily convert an existing Web page to C-HTML. WAP is an international standard language and each major carrier, such as Ericsson, Motorola, Nokia. Since WAP is designed for a handset, users cannot read WAP pages from PCs.

6.4 Impacts to Information Systems

iMode proved that wireless Internet connection is so easy. While a PC takes more than 60 seconds to boot up Windows and connect to the Internet, iMode takes less than 10 second. DoCoMo's 6.5 million subscribers plus competitors' five millions subscribers total 11.5 million. This is almost equivalent to PC Internet users in Japan. Before iMode, Japanese Internet penetration was 13.4% (1998), while in the US it was 37.0%. The latest survey predicts that penetration will be 56.5% and 48.5% in 2001, respectively.

iMode is also changing the Internet business. NTT DoCoMo has become the biggest Internet provider in Japan. AOL took 15 years to gain 24 million subscribers; however, NTT DoCoMo obtained 6.5 million customers with just 15 'months'. If more people access the Internet from cell phones rather than from PC, DoCoMo's portal site would be more attractive than Yahoo.

The market potential for iMode is huge. Japan's market potential is estimated at 360 million units, which is estimated to be three times as much as the entire population by 2010. Japan's wireless market generates $50 billion in revenue. An industry association estimated that the revenues in 2003 would be $100 billion.

6.5 Why is iMode so Successful?

There are more reasons than one. It is to a large extent the fact that NTT-DoCoMo has made it easy for developers to develop iMode websites, that PCs in Japanese homes are not so widespread as in the USA and Europe, so that Japanese people do not use PCs for Internet access as much as in the US or Europe and several other reasons. Here is a list of possible reasons:

- Relatively low street price to consumers for iMode-enabled handsets at point of purchase = low entrance threshold.
- High mobile phone penetration (60 million mobile subscribers).
- Relatively low PC penetration at home.
- iMode uses packet switched system: 'always on' (if the iMode signal reaches the handset etc.), charges according to information accessed not usage time, relatively low fees.
- Efficient micro-billing system via the mobile phone bill. The micro-billing system makes it easy for subscribers to pay for value added, premium sites, and attractive for site owners to sell information to users.
- Up-to-date and efficient marketing.
- E-mail is the 'killer app' as in the initial years of Internet growth.
- Uses cHTML, which makes it easy not only for developers but also for ordinary consumers to develop content. Explosive growth of content.
- AOL-type menu list of partner sites gives users access to a list of selected content on partner sites which are included in the micro-billing system and can sell content and services.

6.5.1 Bandwidth for Downloading Data

The maximum speed for download is 9.6 kbit/sec. This is approximately six times slower than a 64 kbit/sec ISDN connection, but it is sufficient for simple iMode data. Of course this speed makes it impossible to download live movies through iMode. iMode on FOMA (3G) is much quicker: 384 kbit/sec download and 64 kbit/sec upload in best conditions.

6.6 Security on iMode

Mobile commerce (m-commerce) is conducted on iMode including mobile banking and security trading; therefore security is a serious issue.

The security issues on iMode are divided into different sectors:

(1) Security of the radio link between the iMode handset and the cellular base station (this link uses proprietary protocols and encoding controlled by NTT DoCoMo).
(2) Security of the transparent public Internet connection between iMode sites and the handset in the cHTML layer.
(3) Security of private networks on iMode.
(4) Security of private network links between the iMode center and special service providers such as banks.
(5) Password security.

Each of these different security issues needs to be addressed separately. The iMode network and iMode handsets are equipped for SSL (secure socket layer) encrypted transmission, and iMode handsets have unique identifiers allowing similar security to be implemented as on the wired Internet. Mobile banking on iMode and corporate networks will usually use SSL protected encrypted communication.

6.7 iMode 4G

At present the download speed for iMode data is limited to 9.6 kbit/sec which is about six times slower than an ISDN fixed line connection. Recently, with 504i handsets the download data rate was increased three-fold to 28.8 kbps. However, in actual use the data rates are usually slower, especially in crowded areas, or when the network is 'congested'. For third generation mobile (3G, FOMA) data rates are 384 kbps (download) maximum, typically around 200 kbps, and 64 kbps upload since spring 2001. Fourth generation (4G) mobile communications will have higher data transmission rates than 3G. 4G mobile data transmission rates are planned to be up to 20 megabits per second.

NTT-DoCoMo and Hewlett-Packard have announced that the two companies are jointly developing technologies for 4G wireless communications. They have named the technology platform: MOTO-Media.

Initially, DoCoMo planned to introduce 4G services around 2010. Recently, DoCoMo announced plans to introduce 4G services from 2006, i.e. four years earlier than previously planned.

6.7.1 4G Data Rates in Japan

At present, the 2G iMode data rates in Japan are up to 9.6 kbit/sec, but are usually a lot slower, and 28.2 kbps for 504i handsets (since May 2002). For 3G (FOMA) data rates are at present around 200 kbps (download) and 64 kbps (upload) and may in the future be upgraded to 2 Mbit/sec. For 4G data rates up to 20 Megabytes per second are planned. This is about 2000 times faster than present mobile data rates, and about 10 times faster than top transmission rates planned in the final build out of 3G broadband mobile. It is about 10–20 times faster than standard ASDL services, which are being introduced for Internet connections over traditional copper cables at this time.

Of course it is impossible to predict technology developments and the evolution of culture and customer needs. 4G in principle will allow high-quality smooth video transmission.

6.8 Conclusion

Therefore, iMode is a combination of available technologies. Although CDMA and C-HTML are not distinguished technologies, iMode proved that these technologies were enough to satisfy the needs of 6.5 million customers. They want a quick continuous connection to the Internet from anywhere, anytime, at reasonable costs, though transmission speed is lower than their PC Internet connections.

Moreover, iMode is changing the information system of existing businesses. Customers no longer want to call somebody in an office. They need a direct contact to their sales representative and customized information that they can read with their smart phones.

iMode demands that information managers change their system from the CEO's secretary to assistants of sales representatives.

7

UMTS

7.1 Introduction

UMTS stands for 'Universal Mobile Telecommunications System'.

UMTS is one of the major new 'third generation' (3G) mobile communications systems being developed within the framework defined by the ITU and known as IMT-2000. UMTS will play a key role in creating the mass market for high-quality wireless multimedia communications that will exceed two billion users worldwide by the year 2010. This market will be worth over one trillion US dollars to mobile operators over the next 10 years.

UMTS will enable the wireless Information Society, delivering high-value broadband information, commerce and entertainment services to mobile users via fixed, wireless and satellite networks.

UMTS will speed convergence between telecommunications, IT, media and content industries to deliver new services and create fresh revenue-generating opportunities.

UMTS will deliver low-cost, high-capacity mobile communications offering data rates as high as 2 Mbit/sec under stationary conditions with global roaming and other advanced capabilities. The first UMTS services launched commercially in 2001 and over 100 3G licenses have already been awarded.

Experimental UMTS systems are now in field trial with several leading vendors worldwide. One of the benefits of UMTS from the carriers' point of view is that UMTS builds on current investments in second-generation mobile systems. UMTS also has the support of several hundred network operators, manufacturers and equipment vendors worldwide.

The subject of intense worldwide efforts on research and development throughout the last 10 years, UMTS has the support of many major telecommunications operators and manufacturers because it represents a unique opportunity to create

Wireless Data Technologies. Vern A. Dubendorf
© 2003 John Wiley & Sons, Ltd ISBN: 0-470-84949-5

a mass market for highly personalized and user-friendly mobile access to tomorrow's Information Society. UMTS will deliver pictures, graphics, video communications and other wideband information as well as voice and data, direct to people who can be on the move. UMTS builds on the capability of today's mobile technologies (like digital cellular and cordless) by providing increased capacity, data capability and a far greater range of services using an innovative radio access scheme and an enhanced, evolving core network.

The launch of UMTS services ushers in a new, 'open' communications universe, with players from many sectors (including providers of information and entertainment services) coming together harmoniously to deliver new communications services, characterized by mobility and advanced multimedia capabilities. The successful deployment of UMTS will require new technologies, new partnerships and the addressing of many commercial and regulatory issues. The UMTS Forum is at the heart of all these issues, and encourages you to join us as an active participant in making the Information Society of tomorrow a reality.

7.2 What is UMTS?

- UMTS stands for 'Universal Mobile Telecommunications System'.
- UMTS is one of the major new 'third generation' (3G) mobile communications systems being developed within the framework defined by the ITU and known as IMT-2000.
- UMTS will play a key role in creating the mass market for high-quality wireless multimedia communications that will exceed two billion users worldwide by the year 2010. This market will be worth over one trillion US dollars to mobile operators over the next 10 years.
- UMTS will enable the wireless Information Society, delivering high-value broadband information, commerce and entertainment services to mobile users via fixed, wireless and satellite networks.
- UMTS will speed convergence between telecommunications, IT, media and content industries to deliver new services and create fresh revenue-generating opportunities.
- UMTS will deliver low-cost, high-capacity mobile communications offering data rates as high as 2 Mbit/sec under stationary conditions with global roaming and other advanced capabilities.

7.3 A Brief History of UMTS

- The first UMTS services launched commercially in 2001.
- Over 100 3G licenses have already been awarded.

- UMTS experimental systems are now in field trial with several leading vendors worldwide.
- UMTS builds on current investments in second generation mobile systems.
- UMTS has the support of several hundred network operators, manufacturers and equipment vendors worldwide.

UMTS has been the subject of intense research and development throughout the last 10 years and has the support of many major telecommunications operators and manufacturers. UMTS can deliver pictures, graphics, video communications and other wideband information as well as voice and data, direct to people who can be on the move. UMTS builds on the capability of today's mobile technologies (like digital cellular and cordless) by providing increased capacity, data capability and a far greater range of services using an innovative radio access scheme and an enhanced, evolving core network. The successful deployment of UMTS will require new technologies, new partnerships and the addressing of many commercial and regulatory issues.

UMTS provides data speeds of up to 2 Mbps, making portable videophones a reality.

UMTS builds on and extends the capability of today's mobile, cordless and satellite technologies by providing increased capacity, data capability and a far greater range of services using an innovative radio access scheme and an enhanced, evolving core network.

7.4 Spectrum for UMTS

WRC 2000 identified the frequency bands 1885–2025 MHz and 2110–2200 MHz for future IMT-2000 systems, with the bands 1980–2010 MHz and 2170–2200 MHz intended for the satellite part of these future systems.

7.5 Phases Towards the Development of UMTS

Full commercial deployment will be reached through the following main steps:

- Extension of GSM's capability with packet and high speed data operation.
- Pre-UMTS Trial Phase either in subsets of real GSM networks or in isolated packet-based networks.
- Basic deployment phase in 2002, including the incorporation of UTRA base stations into 'live' networks and the launch of satellite-based UMTS services.

Creating a 3G network requires the installation of a large number of 3G masts, a process that had been hampered by public opposition, due to concerns over possible health dangers. In many parts of the country there are local groups fighting the building of mobile phone masts, and there are concerns that this could make it extremely tricky for mobile operators to complete the roll-out of 3G.

7.6 UMTS/3G Industry

The easiest way to describe what a UMTS network is that you take an existing GSM network, add a high speed Internet connected data network, install CDMA base stations that enable higher data rates and more accurate location information, then you add more applications to make it mobile Internet-like, give fancy color screen multimedia terminal to your customers and your UMTS network is ready. You hope that higher data rates will create a new mobile application industry, that will use all the available bandwidth and you find people who are ready to pay for it. And if all goes well customers learn to call and be called by non-human counterparts and companies see advantages in using the UMTS network as a computer-to-computer communication path and the good times are back for everyone.

7.6.1 Cost

On average, an operator needs about $2.5 billion to build a 5000 base station UMTS network and about the same amount to run the organization and attract customers to make it profitable. Add to this the fact that the operator has to pay for the license and financing. This means that an operator has to come up with an innovative business case to get the funding organized and the network on air.

7.7 3G and UMTS Technology

Mobile data communications are evolving quickly because of Internet, intranet, laptops, PDAs and increased requirements of workforce mobility. UMTS will be the commercial convergence of fixed line telephony, mobile, Internet and computer technology. New technologies are required to deliver high-speed location and mobile-terminal-specific content to users. The emergence of new technologies thus provides an opportunity for a similar boom what the computer industry had in the 1980s, and Internet and wireless voice had in the 1990s.

The main IMT-2000 standardization effort was to create a new air interface that would increase frequency usage efficiency. The WCDMA air interface was selected for paired frequency bands (FDD operation) and TDCDMA (TDD operation) for unpaired spectrum. CDMA2000 standard was created to support IS-95 evolution.

The UMTS transport network is required to handle high data traffic. A number of factors were considered when selecting a transport protocol: bandwidth efficiency, quality of service, standardization stability, speech delay sensitivity and the permitted maximum number of concurrent users. In the UMTS network, ATM (Asynchronous Transfer Mode) is defined for the connection between UTRAN and the core network and may also be used within the core network. In addition to the IMT-2000 frame many new standards will be integrated as part of the next-generation mobile systems. Bluetooth and other close range communication protocols and several different operating systems will be used in mobiles. The Internet will come to mobiles with WAP, i-mode and XML protocols. 3G development has helped to start the standardization and development of a large family of technologies.

This section covers some of the core UMTS technologies.

7.8 3G Network Planning

7.8.1 Prerequisite for a 3G Network Design

Designing a cellular network is like doing a crossword puzzle without any instructions. With 3G, you have the added complication of both the operator and customers holding back some of the pieces and planners having to force pieces to fit together because the edges are not a perfect match.

Getting the required information for a network plan is the most crucial part in building a cost effective quality network. Unfortunately, some of the necessary data is confidential and not only that, it's anybody's guess as to what the 3G mobile service mix and usage will be. In an environment where operators need comprehensive designs and redesigns in a very short time frame, there's plenty to get jittery about. Other vendors are rumored to give better coverage with fewer sites, operators claim that other vendor's products are superior and vendor sales people seem to be promising everything to get the deal. On top of that, timetables keep on changing, but of course the network launch date remains fixed. For the vendor, the worst thing is that once you've won the contract, you actually have to build the network you promised!

A lot of different information from various sources is needed for initial network (rollout) plan. Following is a non-exhaustive list of required data.

7.8.2 Operator's Business Plan

This should define what kind of service the operator is planning to provide, how these services will be implemented and how much money is needed for the total rollout. Sometimes this information is public knowledge and sometimes it is a well guarded secret.

7.8.2.1 Technical Section of Business Plan

This should contain the desired coverage, capacity, quality, features, service mix and customer intake plans.

7.8.3 UMTS License Agreement

This usually contains the coverage, capacity and service deployment plans as well as requirements to hire a predetermined amount of employees and perhaps the required amount of domestic goods and services that need to be purchased.

7.8.4 Operators Funding Plan

This should give guidelines of how the rollout should progress. However, usually operators do not want to share this information. Quite a few UMTS networks are vendor financed and such information could help network planners estimate the rollout pace.

7.8.5 Operators Risk Analysis Documents

The risk analysis documents show where bottlenecks will be as well as show the project's critical path. Most often site acquisition is in the critical path, this means that site RF planning will have to compromise some of the desired sites. However, technically the air interface capacity is normally the network limiting capacity factor and so network rollout planning should really be started from there.

7.8.6 Consultant Reports

It has been the practice that operators asked a lot of consultants to do 3G rollout analysis reports. Even though this information gets old quickly, these reports might have some helpful facts.

7.8.7 Government Statistics

Government sources provide statistics of population type and information such as income, distribution of wealth, taxation, spending habits etc., which are useful to estimate future mobile usage in different areas.

All of this information is contained in summary format in the operators request for quote which calls for an estimation of how many base station locations each network vendor thinks is required to provide a network. The operator usually asks vendors to guarantee the level of coverage for a certain load level, using the minimum amount of base stations and cost. Vendors have to commit to these figures even when most of the sites are yet to be acquired and some of the performance parameters will be defined later. Vendors are expected to reply in a very short period of time with limited information, so it is easy to see why network quality is not the biggest consideration in initial planning. There is a tendency for some operators to use this tactic to get the lowest possible initial quote from vendors.

Most experienced network planners can produce an estimated network base station requirement figure with just a few parameters. The most crucial parameters for the initial rollout are:

- Capacity requirements – which is the planned customers and service usage in each area of the network (with BTS site capacity calculation) should be known in order to get the required amount of base stations for capacity.
- Coverage requirements – this is composed of the link budget of how data rate services should be calculated in order to estimate the required base station amount in each network area to get the amount of base stations for coverage.

In each network area we take the larger number of coverage base stations for that area and then add each area together to get a total. To get the final required number of base stations, the following formula can be used:

- Add 10% more quality sites to provide special coverage or a dominant server in difficult or important areas – tunnels, bridges, exhibition and sports venues, shopping centers, airports, big hotels, high rise buildings.
- Add additional 10% more sites to fix holes because not all planned sites can be acquired.
- By this time the sales team will tell you that your plan is 30% too expensive, so you need to cut 30% of your base stations (and 40% of your acquisition budget).

The total will tell you how many sites you need to build an initial 3G network.

7.8.7.1 Link Budget and Coverage

The WCDMA link budget calculations start from the uplink (reverse link) direction. Uplink interference (noise from other mobiles) is normally the limiting factor in CDMA systems.

The starting point of a link budget calculation is to define the required data rate(s) in each network areas and Eb/No (Energy per Bit to Noise power density ratio) targets. Usually the operator predefines these, but simulation tools can be used to tailor the Eb/No. Simulation can be done by creating a uniform base station and a mobile distribution plan with defined service profiles. Almost every UMTS vendor has a simulation tool for operators to test their network plan models.

The next step is to gather vendor-specific data like a BTS output power and a receiver noise figure, defined and used cable systems (the thicker the cable, the more expensive it is to install), used antenna types, usage of intelligent antenna systems in specific areas, possible additional line amplifiers, used diversities (like antenna, polarization, receiver) etc.

Mobile power levels, the chip rate and the process gains are defined by the UMTS standards. Soft handover gain and the thermal noise density are the same in every UMTS system. Both parties also have to agree on propagation models after drive tests.

The link budget gives a cell range and from that cell coverage area can be calculated. Cell coverage overlap parameter is usually missing from the calculation as it increases the cell count drastically. The majority of network planners agree that overlap should be 20–30%, but that relates directly to build cost. After all that, the base station requirements for the each type of area can be calculated.

7.8.7.2 UMTS Capacity Planning

The number of installed transceivers limits the mobile network theoretical capacity.

In CDMA systems interference, accepted and planned quality and grade of service determines the system capacity. CDMA systems have what is know as soft capacity. This complicates the network area capacity estimations. The link budget is used to calculate the maximum allowed path loss and the maximum range for cell. The link budget includes the interference margin, which is the increased noise level caused by a greater load in a cell. So by increasing the cell load, cell coverage area becomes smaller.

System capacity planning can be divided into two parts:

- The first part is to estimate a single transceiver and site capacity. Calculations of how the noise raises as the cell load increases is out of the scope of this page, but in-cell noise, Eb/No requirements, planned data rates, coverage probability, air resources usage activity factor, target interference margin and processing gains are needed to approximate the transceiver and site capacity. Depending on the parameter values, planned transceiver capacity is typically from 400 kbits/s to 700 kbits/s per transceiver.
- The second part of the process is to estimate how many mobile users each cell can serve. Once the cell capacity and subscriber traffic profiles are known, network area base station requirements can be calculated. Estimations can be done in Erlangs per subscriber or kilobits per subscriber. The network vendor normally has simulation tools to test system parameters and verify rough estimations. A lot of data is required for comprehensive network dimensioning; number of subscribers and growth estimations, traffic/user/busy hour/geographic segment and required throughput including service mixes in geographic segments for example.

7.8.7.3 Common Design Guidelines

Upon completion of calculating the coverage and capacity requirements in each geographical area, the greater one of those two values has to be chosen. Requirements should match in each geographical area, but usually that does not happen. To optimize the used resources some readjustments should be made.

If a geographical area is coverage limited then the load on each sector can be reduced until the coverage and capacity requirements match. By reducing the load, this will cut the link budget interference margin and increase Node B count. If an area is capacity limited, transmitter diversity can be added or the amount of transceivers can be increased.

Operators are usually forced to co-locate their 3G base stations with existing sites or select new site locations only on buildings known to be owned by cooperative site owners. This practice places limits on the cell planning options and can result in sacrificing the network quality, but it also helps to build networks faster. Forced co-location needs to be taken into account in initial capacity and coverage planning. All variation to standard configuration may need pilot power, handover, antenna, cable and base station power level modifications.

There are some network areas that need special attention such as very dense urban area (CBD), open spaces, in-building areas, water surroundings and hot spots need to have a well planned approach. Out-of-Cell Interference versus soft-handover cell overlap needs to be considered. Hierarchical systems work with multi-frequency networks, but they do not work with single-frequency systems (like cdma). If multilayered is planned, separate frequencies are needed for different layers.

7.8.8 RAN Planning

Planning the UMTS RAN and core network side is basically selecting the desired network layout, future expansion approach, calculating the required hardware, deciding software features and dimensioning all interfaces.

The Radio Access Network (RAN) has several interfaces that need to be configured and dimensioned. The RAN interfaces that need to be configured:

- Iu: Interface between the RNC and the Core Network (MSC or SGSN).
  - Iucs: Iu circuit switched (voice from/to MSC).
  - Iups: Iu packet switched (data from/to SGSN).
- Iub: Interface between the RNC and the Node B.
- Iur: Interface between two RNCs.

The Node B amount is determined from air interface capacity and coverage calculations and the Node Bs also have to be configured. Remember that hardware configuration is vendor specific. Below is a general list of things that must be considered when configuring Node Bs:

- Call mix of expected traffic
- Type of Node Bs (outdoor vs indoor)
- Amount of low capacity Node Bs
- Required redundancies (e.g. 2N, N + 1)
- Required diversities
- Number of carriers per sector
- Number of sectors per Node B
- Number of users
- Voice and data traffic to be carried
- Node B software features
- Required Node B optional features
- Requirements for special antenna systems
- Requirements for power and transmission systems.

The RNC planning is completed only after the air interface dimensioning and network interfaces planning has been done. Once these are prepared then the bandwidth of each RNC link is known.

The process of RNC dimensioning is to calculate the number of RNCs and configuration of RNCs needed to support the radio access network requirements. Any network side equipment will have the trade-offs in configuration selection. Networks can be designed for maximizing the ease of future expansion or for minimizing the total cost. RNC Hardware configuration is vendor specific. Below is a list of things that normally are considered when dimensioning RNCs:

- RNC capacity and configuration options
- Total CS traffic (Erlangs)
- Total PS traffic (Mbps)
- Total traffic and signaling load
- Total number of Node Bs
- Total number of cells
- Total number of carriers
- Used channel configurations
- RNC software features
- Required RNC optional features
- Type of transmission interfaces
- Expansion possibilities.

7.8.9 Core Network Planning

The planning of the UMTS core network consist of GSN (GPRS Service Node) system design, MSC and registers dimensioning, OMC dimensioning, Core network interface dimensioning.

The Core Network has several interfaces that need to be configured and dimensioned:

- Gn: Interface between SGSN and GGSN.
- Gi: Interface between GGSN and external packet data network.

Other interfaces are between MSCs, to PSTN, HLR, AUC, EIR, SMS, Billing Center, Voice Mail, OMC, WAP and Multi Media Servers and other network elements.

The main inputs to dimensioning of GSN system are similar to that which is required in the air interface design. Again, hardware configuration is vendor specific:

- Number of Subscribers
- Number of PDP Contexts
- Service Activation Rate
- Peak Traffic amount and overheads (bits/s or packets/s)
- Number of required links
- Number of RNC in served area.

The Core Network hardware configuration is also vendor specific. Network vendors have very extensive documentation on how to design all aspects of core network starting from the room environmental requirements up to the post integration system quality audit.

- Number of subscribers
- Average call lengths
- Call mix
- Total CS traffic (Erlangs)
- System features
- Interconnection to other equipment
- MSC software features
- Required redundancies (e.g. 2N, N + 1)
- Total traffic and signaling loads

- Iu-cs and other interface dimensioning
- Type of transmission interfaces
- Expansion possibilities
- MSC capacity and configuration options
- Most cost effective deployment method
- Number of RNC in served area.

The goal of the network planning is not only to define the initial network roll-out targets, but also to provide moving targets to the continuous process that takes the whole life time of the network. Before the 3G network is launched, all the work is focused on estimating how the network should look. After the network launch, customer intake and behavior will decide the network development direction.

7.8.10 UMTS Security

The security functions of UMTS are based on what was implemented in GSM. Some of the security functions have been added and some existing ones have been improved. The encryption algorithm is somewhat stronger and is included in base station (NODE-B) to radio network controller (RNC) interface, the application of authentication algorithms is also stricter and subscriber confidentially is tighter.

The main security elements that are from GSM:

- Authentication of subscribers
- Subscriber identity confidentially
- Subscriber Identity Module (SIM) to be removable from terminal hardware
- Radio interface encryption.

Additional UMTS security features include:

- Security against using false base stations with mutual authentication.
- Encryption extended from air interface only to include Node-B to RNC connection.
- Security data in the network will be protected in data storages and while transmitting ciphering keys and authentication data in the system.
- Mechanism for upgrading security features.

Core network traffic between RNCs, MSCs and other networks is not ciphered and operators can implement protections for their core network transmission links, however this is very unlikely to happen. MSCs will have by design a lawful interception capabilities and access to Call Data Records (SDR), so all switches will have to have security measures against unlawful access.

The UMTS specification has five security feature groups:

- Network access security: this is a set of security features that provide users with secure access to 3G services, and which in particular protect against attacks on the (radio) access link.
- Network domain security: this set of security features that enables nodes in the provider domain to securely exchange signaling data, and protects against attacks on the wireline network.
- User domain security: this is a set of security features that secures access to mobile stations.
- Application domain security: this is a set of security features that enables applications in the user and in the provider domain to securely exchange messages.
- Visibility and configurability of security: this is a set of features that enables the user to inform himself of whether a security feature is in operation or not and whether the use and provision of services should depend on the security feature.

The UMTS specification also has user identity confidentiality security features which are:

- User identity confidentiality: the property that the permanent user identity (IMSI) of a user to whom a services is delivered cannot be eavesdropped on the radio access link.
- User location confidentiality: the property that the presence or the arrival of a user in a certain area cannot be determined by eavesdropping on the radio access link.
- User untraceability: the property that an intruder cannot deduce whether different services are delivered to the same user by eavesdropping on the radio access link.

7.8.11 3G and LAN Data Speeds

Figure 7.1 below contains the theoretical maximum data speeds of 2G, 2.5G, 3G and beyond compared to LAN data speeds.

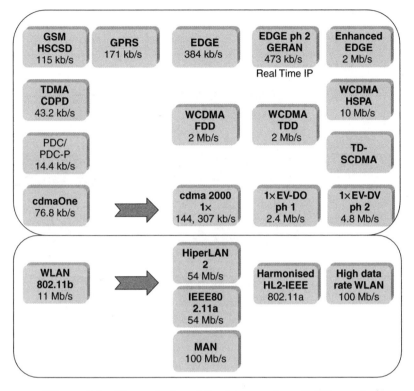

Figure 7.1 Theoretical mobile data speeds vs LAN speeds

7.8.12 3G Frequencies

Refer to Figure 7.2, for a high level view of the UMTS frequencies.

1920–1980 and **2110–2170** MHz Frequency Division Duplex (FDD, W-CDMA) Paired uplink and downlink, channel spacing is 5 MHz and raster is 200 kHz. An Operator needs three to four channels (2×15 MHz or 2×20 MHz) to be able to build a high-speed, high-capacity network.

1900–1920 and **2010–2025** MHz Time Division Duplex (TDD, TD/CDMA) Unpaired, channel spacing is 5 MHz and raster is 200 kHz. Tx and Rx are not separated in frequency.

1980–2010 and **2170–2200** MHz Satellite uplink and downlink.

The carrier frequencies are designated by a UMTS Absolute Radio Frequency Number (UARFN). The general formula relating frequency to UARFN is:

$$\text{UARFN} = 5 * (\text{frequency in MHz})$$

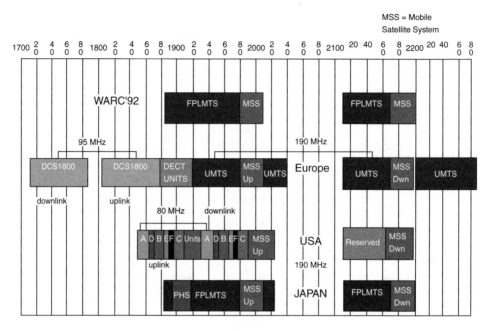

Figure 7.2 UMTS frequencies

7.9 Owners of the 3G Networks

Country	Operator	Owners
Australia	Telstra	51% Australian Government, 49% publicly owned
	Optus	SingTel: publicly owned
	Vodafone	100% Vodafone
	Hutchison	80.1% Hutchison Telecom, 19.9% Telecom New Zealand
	3G Investments	Qualcomm
	CKW Wireless	Arraycom
Austria	Mobilkom	100% Telekom Austria AG (Telekom Austria bought back Telecom Italia's 25% share in June 2002; reported 07/06/02)

Country	Operator	Owners
	Connect	50.1% German E.ON, 17.45% Norwegian Telenor, 17.45% British Orange, 15% Danish Tele Danmark
	Max.mobile	100% T-Mobile International AG
	Tele.ring	100% Western Wireless International
	3G Mobile	100% Telefónica Móviles
	Hutchison 3G	100% Hutchison Whampoa
Belgium	Proximus	75% Belgacom, 25% Vodafone
	KPN Mobile 3G	100% KPN
	Mobistar NV	50.58% WSB (France Telecom), 21.44% Publicly owned, 10 other shareholders
Czech Republic	RadioMobil	39.2%, Ceské radiokomunikace, 60.8% CMobil (92.136% T-Mobile, 7,164% STET and 0.7% PVT)
	Eurotel	51% CESKÝ Telecom, Atlantic West (Verizon & AT&T Wireless)
Denmark	HI3G Denmark	60% Hutchison Whampoa, 40% Investor AB
	TDC Mobile Int	Danish National Telecom Agency, formerly Tele Danmark, 41.6% SBC Communication Inc.
	Telia Mobile	Telia Sweden (70.6% State owned)
	Orange	54% France Telecom, 14% Banestyrelsen, 32%, five other institutions
Finland	Sonera	52.8% Finnish State, publicly owned
	Radiolinja	Helsingin Puhelin

(continued overleaf)

Country	Operator	Owners
	Telia	Telia Sweden
	Suomen Kolmegeen	41 local Finnet-group companies and NetCom Ab
Finland, Åland	Ålands Mobiltelefon	
	Song Networks	
France	SFR	Cegetel: 44% Vivendi, 26% BT Group, 15% SBC Communications, 15% Vodafone
	France Telecom	55.5% France State, 32.6% publicly owned, 4.3% Vodafone, 4.2% Treasure shares, 1.8% Deutsche Telecom, 1.6% Sita
	Bouygues	15.2% SCDM, 53.9% publicly owned, 10.1% Artémis, employees and institutions
Germany	T-Mobil	Deutsche Telecom: 30.92% German State, 12.13% KfW, 56.95% publicly owned
	Vodafone D2 (ex Mannesmann)	65.23% Vodafone AG, 34.77% Vodafone Europe GmbH & Co. KG
	E-Plus	100% KPN
	Viag Interkom	45% Viag, 45% BT, 10% Telenor
	Mobilcom	40% Gerhard Schmid, 28.5% France Telecom, 31.5% publicly owned
	Group 3G	57.4% Telefonica, 42.6% Sonera
Greece	CosmOTE	58.98% OTE Telecoms, 18% Telenor, 7.27% WR Com Enterprises, 15.75% publicly owned
	STET Hellas	62.804% TIM, 17.453% NYNEX, 5.051 Others, 14.692% publicly owned

Country	Operator	Owners
	Panafon	51.88% Vodafone Group, 10.85% France Telecom, 9.43% Intracom, 27.84% publicly owned
Hong Kong	Hong Kong CSL	60% Telstra Corporation Limited, 40% Pacific Century CyberWorks Limited
	Hutchison 3G	74.63% Hutchison Whampoa Limited, 25.37% NTT DoCoMo
	SmarTone 3G	SmarTone: 29.36% Sun Hung Kai Properties, 20.75% BT, publicly owned
	SUNDAY 3G	SUNDAY Communications: 46.2% Distacom Communications Limited, 11.5% USI Holdings Limited, publicly owned
Israel	Cellcom:	BellSouth Corp, Brazil's Safra group and Israel's Discount Investment Corp
	Partner Communications	35% Hutchison Whampoa, 15.51% Matav, 12.37% Elbit, 12.37% Tapuz and 24.75% publicly owned
	Pele-phone	Bezeq Israel Telecom and Shamrock Holdings of Roy Disney
Italy	Wind	51% Enel, 24.5% France Telecom and 24.5% Deutsche Telekom
	Omnitel	76.86% Vodafone, 23.14% Verizon
	TIM	56.1% Telecom Italia, publicly owned
	IPSE	45.59% Telefonica, 12.55% Sonera, 12% Atlanet, 10% Banca di Roma, 5% Xfera, 5% Edison-Flack, 4.8% Goldenegg, 5% others

(continued overleaf)

Country	Operator	Owners
	H3G (ex Andala)	88.2% Hutchison, 1.8% Cir, San Paolo IMI 5.6%, BMI 2.3%, Hdp 1.1%, Gemina 0.6% and Tiscali 0.4%
Japan	NTT DoCoMo	64.1% Nippon Telegraph and Telephone, publicly owned
	DDI	KDDI: publicly owned
	J-Phone	69.7% Vodafone, Japan Telecom (Vodafone: 39.67% Japan Telecom: 45.08%. Vodafone is the major shareholder of Japan Telecom, with 66.7% of outstanding shares)
Liechtenstein	VIAG	VIAG Interkom GmbH & Co., Germany
	Tele2	Tele2 AB Sweden
Luxembourg	EPT	publicly owned
	Orange	France Telecom
	Tango S.A.	100% Tele2 AB Sweden
Monaco	Monaco Telecom	55% Vivendi, 45% Monaco State
The Netherlands	Libertel	70% Vodafone, publicly owned
	KPN	85% Koninklijke KPN, 15% NTT DoCoMo
	Dutchtone	100% Orange SA (France Telecom)
	Telfort	100% mmO2 plc: BT, publicly listed
	3G Blue	50% Deutsche Telecom, Ben 50% (Ben: 70.6% Belgacom, 29.4% Tele Danmark)
New Zealand	Telstra Clear	62% Telstra Corporation, 38% Austar United Communications

Country	Operator	Owners
	Telecom NZ	100% Telecom New Zealand (Publicly owned), Hutchison Whampoa has 19.9% option to subscribe
	Vodafone NZ	100% Vodafone
	Econet Wireless	Te Huarahi Tika Trust
Norway	Telenor	77.66% Norwegian Government, 1.79% Folketrygdfondet. Publicly owned
	Netcom	Telia AB Sweden
	Tele 2	NetCom AB Sweden
Poland	Centretel	66%, Telekomunikacja Polska S.A., 34% France Telecom
	PTC	26% Vivendi, 49% T-Mobil
	Polkomtel	19.6% Vodafone, 19.6% TeleDanmark, 61.5% seven Polish firms
Portugal	TMN	Portugal Telecom
	Telecel	50.9% Vodafone, publicly owned
	Optimus:	45% Sonae, 20% France Telecom, 25.49% Thorn Finance
	Oni Way	55% Oni SGPS, 20% Norway's Telenor, Iberdrola, Media Capital, Efacec
Singapore	MobileOne	35% Keppel Group, 35% Singapore Press Holdings, 30% Great Eastern Telecommunications (GET: 51% C&W, 49% PCCW)
	Starhub Mobile	50.47% Singapore Technologies Telemedia, 14.51% NTT, 14.07% Media Corporation of Singapore, 11.87% BT, 9.08% Singapore Press Holdings

(*continued overleaf*)

Country	Operator	Owners
	STM	Singapore Telecom (78% state, publicly owned)
Slovenia	Mobitel	100% Telecom Slovenia
South Korea	Korea Telecom	KTICOM: KT group, 500 shareholder companies
	SK Telecom	SK Telecom 3G: 48.5% SK Telecom, 12% Pohang Iron & Steel, Publicly owned
	LG Telecom	35.6% LG Electronics, 16.6% BT, Publicly owned
Spain	Telefónica Móviles	Telefónica, publicly listed
	Xfera	31% Vivendi & FCC, 34% ACS & Sonera, 6% Acesa, 7% Mercapital, 7% CF Alba 3% JP Morgan, 5% Cajas De Ahorro
	Airtel Movil	93.8% Vodafone, 6.2% Acciona S.A
	Amena	97.9% Auna, 2.1% Caixa de Catalunya (Auna 'brings together eight companies: Retevisión, Amena, eresMas, Madritel, Menta, Able, Supercable Andalucia and Telecom Canarias')
Sweden	Europolitan	71.1% Vodafone, 28.8% Publicly owned
	HI3G	60% Hutchison Whampoa, 40% Investor AB
	Orange	85% Orange SA (France Telecom),10% Skanska AB, 3% NTL Inc, 2% Schibsted ASA
	Tele 2	Telia and Tele2 (Netcom AB) in a 50/50 joint venture Svenska UMTS
Switzerland	Swisscom	25% Vodafone, Publicly listed

Country	Operator	Owners
	Sunrise (Diax)	78.72% Tele Danmark, 16.99% d holding, 2.34% SBB, 1.95% UBS
	Orange	99.9% Orange UK (France Telecom), 0.1% Banque Cantonale Vaudoise
	Team 3G	100% Telefónica
Taiwan	Taiwan PCS Network	Taiwan Paging Network, the China Development Industrial Bank, auto manufacturer Yue Loong Motor Co and US-based monitor firm ViewSonic Corp
	Taiwan Cellular Corp	Pacific Electric Wire & Cable Co., Ltd., Fubon Group, Evergreen Group, Acer Group, Continental Engineering Corp., Yageo Co., Verizon Communications
	Chunghwa Telecom	95.34% Ministry of Transportation and Communications, publicly listed
	Yuan-Ze Telecom	Yuan-Ze Telecommunications is a subsidiary of Far EasTone, which is a JV formed by the Far Eastern Group and AT&T Wireless
	APBW	Main Share holder is China Rebar Co. Ltd
UK	Hutchison (ex TIW license)	65% Hutchison Whampoa, 20% NTT DoCoMo, 15% KPN Mobile
	Vodafone	100% Vodafone (publicly listed)
	BT	100% BT (publicly listed)
	T-Mobile (One2One)	100% Deutsche Telecom (publicly listed)
	Orange	France Telecom, publicly listed
UK, Isle of Man	Manx Telecom	100% BT

8

Wireless Data Networks

Almost all 'Wireless' specialists will agree that the lack of standards was one of the main factors that held up the progress of WLANs. In 1997 the IEEE adopted the first Wireless LAN (WLAN) standard, IEEE Std 802.11-1997. There are three wireless LAN (WLAN) types, each of which uses a different part of the electromagnetic spectrum: Infrared, Microwave, and Spread Spectrum Technology (SST). Each solution has its unique advantages and disadvantages associated with the nature of its electromagnetic spectrum frequency. In 1999 a revision was made to this standard. This IEEE standard defines a medium access control (MAC) sublayer, MAC management protocols and services, and also three physical layers. The three physical (PHY) layers are comprised of a direct sequence spread spectrum (DSSS) radio, a frequency hopping spread spectrum (FHSS) radio, both of which operate in the 2.4 GHz band, and an infrared (IR) baseband. The 802.11 standard describes these layers as providing operations of 1 and 2 Mbps.

The spread spectrum technology is the result of secure transmission methods that were developed by the military during World War II.

With the technology still hampered by slow speeds, the IEEE 802.11 Working Groups are continuing to work on new standards that will give better results. The first is IEEE Standard 802.11a, which is an orthogonal frequency domain multiplexing (OFDM) radio in the UNI bands that have the ability to provide up to 54 Mbps data rates. The second is the IEEE Standard 802.11b which is an extension to the DSSS PSY in the 2.4 GHz bands that delivers data rates up to 11 Mbps. Each of these is an addition to the PHY layer.

After 10 years of discussions, the final approval of the IEEE 802.11 specification for wireless networking came on June 26, 1997. Developed by the Institute of Electrical and Electronics Engineers (IEEE), it can be compared to the 802.3

Wireless Data Technologies. Vern A. Dubendorf
© 2003 John Wiley & Sons, Ltd ISBN: 0-470-84949-5

standard for Ethernet wired LANs. The Physical Layer under 802.11 includes three alternatives covering all the usual forms of WLAN:

- Diffused Infrared (DFIR).
- Direct Sequence Spread Spectrum (DSSS).
- Frequency Hopping Spread Spectrum (FHSS).

Both radio frequency spread spectrum specifications are in the 2.4 GHz band. The 2.4 GHz band was chosen because it is available for unlicensed operation worldwide and because it is possible to build low-cost, low-power radios in this frequency range that operate at LAN speeds. Spread spectrum and low power are requirements to allow unlicensed operation and to avoid interfering with other types of devices that may use the 2.4 GHz band.

8.1 Data Networks and Internetworking

In the same way that this book provides a foundation for understanding wireless technologies, this section builds a foundation for understanding Data Networks and Internetworking which is needed in order to have a full appreciation for Wireless Data Networks. Topics in this chapter will include flow control, error checking, and multiplexing, however this sections' focus is mainly on mapping the Open System Interconnection (OSI) model to networking/internetworking functions, and also on summarizing the general nature of addressing schemes within the context of the OSI model. The OSI model represents the building blocks for internetworks regardless of whether those internetworks are wireless or wired. Understanding the conceptual model will help you understand the complex pieces that make up an internetwork.

8.1.1 What is an Internetwork?

An *internetwork* is a collection of individual networks, wired or wireless that are connected by intermediate networking devices. This internetwork functions as a single large network. Internetworking refers to the industry, products, and procedures that meet the challenge of creating and administering internetworks.

The first networks were time-sharing networks that used mainframes and attached terminals. *Local-area networks* (LANs) evolved around the PC revolution. LANs made it possible for multiple users in a relatively small geographical area to exchange files and messages and also to access shared resources such as file servers and printers. *Wide-area networks* (WANs) interconnect LANs with geographically dispersed users to create connectivity. Some of the technologies used for connecting LANs include T1, T3, ATM, ISDN, ADSL, Frame Relay, wireless or radio links, and others. New methods of connecting dispersed LANs are appearing everyday. Today, high-speed LANs and switched internetworks are becoming widely used because they operate at very high speeds and support such high-bandwidth applications as multimedia and videoconferencing.

Internetworking evolved as a solution to three key problems: isolated LANs, duplication of resources, and a lack of network management.

Isolated LANs made electronic communication between different offices or departments impossible. Duplication of resources meant that the same hardware and software had to be supplied to each office or department, as did separate support staff. This lack of network management meant that no centralized method of managing and troubleshooting networks existed.

It is not an easy task implementing a functional internetwork. A number of challenges must be faced such as in the areas of connectivity, reliability, network management, and flexibility. Each area is key in establishing an efficient and effective internetwork.

One of the great challenges when connecting various systems is to support communication among disparate technologies. Different sites, for example, may use different types of media operating at varying speeds, or may even include different types of systems that need to communicate such as Microsoft Windows, Novell, Unix, or even AIX.

Because companies rely heavily on data, communication internetworks must provide a certain level of reliability. Many large internetworks include redundancy to allow for communication even when problems occur using protocols such as Cisco Systems HSRP (Hot Standby Router Protocol) or the generic version VRRP (Virtual Redundant Router Protocol).

Network management must provide centralized support and troubleshooting capabilities in an internetwork. Configuration, security, performance, and other issues must be adequately addressed for the internetwork to function properly and smoothly. Also essential within an internetwork is Security. The majority of people think of network security as the act of protecting the private network from outside attacks, however it is just as important to protect the network from internal attacks because most security breaches come from inside. Networks must also be secured so that the internal network cannot be used as a tool to attack other external sites.

Early in the year 2000, many major web sites were the victims of distributed denial of service (DDOS) attacks. These attacks were possible because a great number of private networks currently connected with the Internet were not properly secured. These private networks were used as tools for the attackers.

Nothing in this world is stagnant. We are constantly evolving and changing and so is anything we touch. For this reason, internetworks must be flexible enough to change with new demands.

8.1.2 Open System Interconnection Reference Model

The *Open System Interconnection* (OSI) *reference model* describes how information from a software application in one computer moves through a network medium to a software application in another computer. The OSI reference model is a conceptual model composed of seven layers, each specifying particular network functions. This model was developed by the International Organization for Standardization (ISO) in 1984, and it is now considered the primary architectural model for intercomputer communications.

The OSI model divides the tasks involved with moving information between networked computers into seven smaller, more manageable task groups. A task or group of tasks is then assigned to each of the seven OSI layers. Each layer is reasonably self-contained so that the tasks assigned to each layer can be implemented independently. This enables the solutions offered by one layer to be updated without adversely affecting the other layers. The following list details the seven layers of the Open System Interconnection (OSI) reference model. My wife teaches an easy way to remember the seven layers using the sentence, 'All people should tell no lies period.' The beginning letter of each word corresponds to a layer.

- Layer 7 – Application – All
- Layer 6 – Presentation – People
- Layer 5 – Session – Should
- Layer 4 – Transport – Tell
- Layer 3 – Network – No
- Layer 2 – Data link – Lies
- Layer 1 – Physical – Period.

The seven layers of the OSI reference model can be divided into two categories:

- upper layers;
- and lower layers.

The *upper layers* of the OSI model deal with application issues and are normally found implemented only in software. The highest layer, which is the application

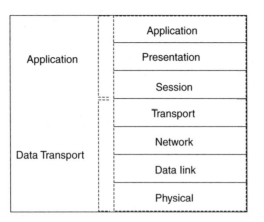

Figure 8.1   Illustration of the division between the upper and lower OSI layers

layer, is also the closest layer to the end user. Both users and application layer processes interact with software applications that contain a communications component. The term upper layer is sometimes used to refer to any layer above another layer in the OSI model. See Figure 8.1 for an illustration of the division between the upper and Lower OSI layers.

The *lower layers* of the OSI model handle all of the data transport issues. The physical layer and the data link layer are implemented in both hardware and software. The lowest layer, which is the physical layer, is closest to the physical network medium (the network cabling, for example) and is responsible for actually placing information on the medium.

8.1.3 OSI Protocols

Even though the OSI model itself provides a conceptual framework for communication between computers we must remember that the model itself is not a method of communication. The communication is actually made possible by using communication protocols. In the context of data networking, a *protocol* is a formal set of rules and conventions that governs how computers exchange information over a network medium. A protocol implements the functions of one or more of the OSI layers.

A wide variety of communication protocols exist. Some of these protocols include Wireless protocols, LAN protocols, WAN protocols, network protocols, and routing protocols. *LAN protocols* operate at the physical and data link layers of the OSI model and define communication over the various LAN media. *WAN protocols* operate at the lowest three layers of the OSI model and define communication over the various wide-area media. *Routing protocols* are network layer protocols that are responsible for exchanging information between routers so that the routers can select the proper path for network traffic. *Network protocols* are the various upper-layer protocols that exist in a given protocol suite. *Wireless Protocols* can operate in all LAN, WAN, Routing and Network protocol groups. Many protocols rely on others for operation. An example of this is that many routing protocols use network protocols to exchange information between routers. This concept of building upon the layers already in existence is the foundation of the OSI model.

8.1.4 OSI Model and Communication Between Systems

Information being transferred from a software application in one computer system to a software application in another computer is required to pass through the OSI layers. An example of this would be if a software application in System A has information to transmit to a software application in System B then the application program in System A will pass its information to the application layer (Layer 7) of System A. The application layer then passes the information to the presentation layer (Layer 6), which relays the data to the session layer (Layer 5), and so on down to the physical layer (Layer 1). At the physical layer, the information is placed on the physical network medium and is sent across the medium to System B. The physical layer of System B removes the information from the physical medium, and then its physical layer passes the information up to the data link layer (Layer 2), which passes it to the network layer (Layer 3), and so on, until it reaches the application layer (Layer 7) of System B. Finally, the application layer of System B passes the information to the recipient application program to complete the communication process.

8.1.4.1 Interaction Between OSI Model Layers

A given layer in the OSI model normally communicates with three other of the OSI layers. These other layers are: the layer directly above it, the layer directly below it, and its peer layer in other networked computer systems. The data link layer in System A, as an example, communicates with the network layer of System A, the physical layer of System A, and the data link layer in System B.

8.1.4.2 OSI Layer Services

One OSI layer communicates with another layer to make use of the services provided by the second layer. The services provided by adjacent layers help a given OSI layer communicate with its peer layer in other computer systems. Three basic elements are involved in layer services: the service user, the service provider, and the service access point (SAP).

In this context, the *service user* is the OSI layer that requests services from an adjacent OSI layer. The *service provider* is the OSI layer that provides services to service users. OSI layers can provide services to multiple service users. The SAP is a conceptual location at which one OSI layer can request the services of another OSI layer.

The seven OSI layers use various forms of control information to communicate with their peer layers in other computer systems. This *control information* consists of specific requests and instructions that are exchanged between peer OSI layers.

Control information typically takes one of two forms: headers and trailers.

- *Headers* are prepended to data that has been passed down from upper layers.
- *Trailers* are appended to data that has been passed down from upper layers.

The OSI layer is not required to attach a header or a trailer to data from upper layers. Headers, trailers, and data are relative concepts, depending on the layer that analyzes the information unit. At the network layer, for example, an information unit consists of a Layer 3 header and data. At the data link layer, however, all the information passed down by the network layer (the Layer 3 header and the data) is treated as data. This means that the data portion of an information unit at a given OSI layer potentially can contain headers, trailers, and data from all the higher layers. This is known as *encapsulation*.

The information exchange process occurs between peer OSI layers. Each layer in the source system adds control information to data, and each layer in the destination system analyzes and removes the control information from that data.

For the following description, refer to Figure 8.2. If System A has data from a software application to send to System B, the data is passed to the application layer. The application layer in System A then communicates any control information required by the application layer in System B by prepending a header to the data. The resulting information unit (a header and the data) is passed to the presentation layer, which prepends its own header containing control information intended for the presentation layer in System B. The information unit grows in

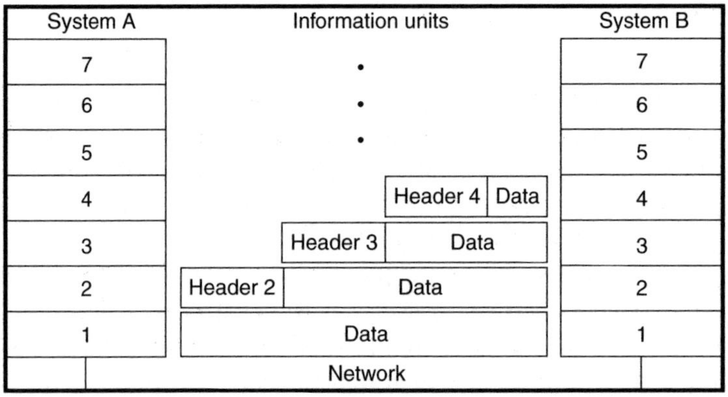

System A	Information units	System B
7	•	7
6	•	6
5	•	5
4	Header 4 | Data	4
3	Header 3 | Data	3
2	Header 2 | Data	2
1	Data	1
	Network	

Figure 8.2 Headers and data can be encapsulated during information exchange

size as each layer prepends its own header (and, in some cases, a trailer) that contains control information to be used by its peer layer in System B. At the physical layer, the entire information unit is placed onto the network medium.

The physical layer in System B receives the information unit and passes it to the data link layer. The data link layer in System B then reads the control information contained in the header prepended by the data link layer in System A. The header is then removed, and the remainder of the information unit is passed to the network layer. Each layer performs the same actions: the layer reads the header from its peer layer, strips it off, and passes the remaining information unit to the next highest layer. After the application layer performs these actions, the data is passed to the recipient software application in System B, in exactly the form in which it was transmitted by the application in System A.

8.2 The OSI Layers

In this section I want to look deeper at the seven layers of the OSI model so that we have a better understanding of their workings.

8.2.1 The Physical Layer – OSI Layer 1

The physical layer defines the electrical, mechanical, procedural, and functional specifications for activating, maintaining, and deactivating the physical link between communicating network systems.

Physical layer specifications define characteristics such as voltage levels, timing of voltage changes, physical data rates, maximum transmission distances, and physical connectors. Physical layer implementations can be categorized as either LAN or WAN specifications.

8.2.2 The Link Layer – OSI Layer 2

The data link layer provides a reliable transit of data across a physical network link. Different data link layer specifications define different network and protocol characteristics. These include physical addressing, network topology, error notification, sequencing of frames, as well as flow control.

Physical addressing, which should not be confused with network addressing, defines how devices are addressed at the data link layer. Network topology consists of the data link layer specifications that often define how devices are to be physically connected, such as in a bus, star, wireless or a ring topology. Error notification alerts upper-layer protocols that a transmission error has occurred, and the sequencing of data frames reorder the frames that are transmitted out of sequence. Finally, flow control moderates the transmission of data so that the receiving device is not overwhelmed with more traffic than it can handle at one time.

The Institute of Electrical and Electronics Engineers (IEEE) has subdivided the data link layer into two sublayers:

(1) Logical Link Control (LLC) and
(2) Media Access Control (MAC).

Communications between devices are managed by the *Logical Link Control* (LLC) sublayer of the data link layer, which supports both connectionless and connection-oriented services, used by higher-layer protocols. IEEE 802.2 defines a number of the fields in the data link layer frames that enable multiple higher-layer protocols to share a single physical data link. The *Media Access Control* (MAC) sublayer of the data link layer manages protocol access to the physical network medium. MAC addresses are defined by the IEEE MAC specification and these enable multiple devices to uniquely identify one another at the data link layer.

8.2.3 The Network Layer – OSI Layer 3

The network layer defines the network address, which differs from the MAC address.

Some network layer implementations, such as the Internet Protocol (IP), define network addresses in a way in which route selection can be determined systematically by comparing the source network address with the destination network address and applying the subnet mask. Since this layer defines the logical network layout, routers can use this layer to determine how to forward packets. Because of this, much of the design and configuration work for internetworks happens at Layer 3, the network layer.

8.2.4 The Transport Layer – OSI Layer 4

Layer 4, the transport layer, accepts data from the session layer and segments the data for transport across the network.

Generally, the transport layer is responsible for making sure that the data is delivered error-free and in the proper sequence. Generally, Flow control occurs here at the transport layer.

Flow control manages data transmission between devices so that the transmitting device does not send any more data than the receiving device can process at a given time. Multiplexing enables data from several applications to be transmitted onto a single physical link. Virtual circuits are established, maintained, and terminated by the transport layer. Error checking involves creating various mechanisms for detecting transmission errors, while error recovery involves acting, such as requesting that data be retransmitted, to resolve any errors that occur.

The transport protocols used on the Internet are TCP and UDP.

8.2.5 The Session Layer – OSI Layer 5

Layer 5, which is the session layer, establishes, manages, and terminates communication sessions.

Communication sessions consist of service requests and service responses that occur between applications located in different network devices. The coordination of these requests and responses are handled by protocols implemented at the session layer. Some examples of session-layer implementations include Zone Information Protocol (ZIP), the AppleTalk protocol that coordinates the name binding process; and Session Control Protocol (SCP), which is the DECnet Phase IV session layer protocol.

8.2.6 The Presentation Layer – OSI Layer 6

The presentation layer provides a variety of coding and conversion functions that are applied to application layer data. These functions ensure that information sent from the application layer of one system would be readable by the application layer of another system.

Examples of the presentation layer coding and conversion schemes include common data representation formats, conversion of character representation formats, common data compression schemes, and common data encryption schemes.

Common data representation formats, or the use of standard image, sound, and video formats, enable the interchange of application data between different types of computer systems. Using different text and data representations, such as EBCDIC and ASCII, uses conversion schemes to exchange information with systems. Standard data compression schemes enable data that is compressed at the source device to be properly decompressed at the destination. Standard data encryption schemes enable data encrypted at the source device to be properly deciphered at the destination.

Presentation layer implementations are not typically associated with a particular protocol stack. More commonly known standards for video include QuickTime and Motion Picture Experts Group (MPEG). QuickTime is an Apple Computer specification for video and audio, and MPEG is a standard for video compression and coding.

Among the most commonly known graphic image formats are Graphics Interchange Format (GIF), Joint Photographic Experts Group (JPEG), and Tagged Image File Format (TIFF). GIF is a standard for compressing and coding graphic images. JPEG is another compression and coding standard for graphic images, and TIFF is a standard coding format for graphic images.

8.2.7 The Application Layer – OSI Layer 7

Layer 7, the application layer, is the OSI layer closest to the end user, which means that both the OSI application layer and the user interact directly with the software application.

This application layer interacts with software applications that implement a communicating component. These types of application programs fall outside the scope of the OSI model. Application layer functions usually include identifying communication partners, determining resource availability, and synchronizing communication.

During communication partner identification process, the application layer determines the identity and availability of communication partners for an application with data to transmit.

During resource availability determination, the application layer needs to decide whether sufficient network resources for the requested communication exist. In synchronizing communication, all communication between applications requires cooperation that is managed by the application layer.

Some examples of application layer implementations include Telnet, File Transfer Protocol (FTP), and Simple Mail Transfer Protocol (SMTP).

8.3 ISO Hierarchy of Networks

Normally, larger networks are organized as hierarchies. A hierarchical organization allows for various advantages such as ease of management, flexibility, and a reduction in unnecessary traffic. The International Organization for Standardization (ISO) has adopted a number of terminology conventions for addressing a networks entity. Key terms defined in this section include end system (ES), intermediate system (IS), area, and autonomous system (AS).

An *ES* is a network device that *does not* perform routing or other traffic forwarding functions. Normally, ESs include devices such as terminals, personal computers, and printers.

An *IS* is a network device that does perform routing or other traffic-forwarding functions. Usually ISs include such devices as routers, switches, and bridges. There are two types of existing IS networks, these are:

- intradomain IS and
- interdomain IS.

An intradomain IS communicates within a single autonomous system and a interdomain IS communicates within and between autonomous systems.

An *area* is a logical group of network segments and their attached devices. Areas are subdivisions of autonomous systems (ASs). An AS is a collection of networks under a common administration that share a common routing strategy. Autonomous systems are subdivided into areas, and an AS is sometimes called a domain.

8.4 Internetwork Addressing

Internetwork addresses identify devices separately or as members of a group.

Addressing schemes vary depending on the protocol family and the OSI layer. Three types of internetwork addresses are commonly used:

- Data link layer addresses
- Media Access Control (MAC) addresses
- Network layer addresses.

8.4.1 Data Link Layer Addresses

A *data link layer address* uniquely identifies each physical network connection of a network device. Data-link addresses are sometimes referred to as *physical* or *hardware addresses*. Data-link addresses are usually found within a flat address space and have a pre-established and typically fixed relationship to a specific device.

End systems normally only have one physical network connection and therefore have only one data-link address. Routers and other internetworking devices usually have multiple physical network connections and therefore have multiple data-link addresses.

8.4.2 MAC Addresses

Media Access Control (MAC) addresses consist of a subset of data link layer addresses. MAC addresses identify network entities in LANs that implement the IEEE MAC addresses of the data link layer. As with most data-link addresses, MAC addresses are unique for each LAN interface.

MAC addresses are 48 bits long and are expressed as 12 hexadecimal digits. The first six hexadecimal digits, which are administered by the IEEE, identify the manufacturer or vendor and comprise the Organizationally Unique Identifier (OUI). The last six hexadecimal digits comprise the interface serial number, or another value administered by the specific vendor. MAC addresses sometimes are called *burned-in addresses* (BIAs) because they are burned into read-only memory (ROM) and are copied into random-access memory (RAM) when the interface card initializes.

8.4.3 Mapping Addresses

Internetworks generally use network addresses to route traffic around the network, hence there is a need to map network addresses to MAC addresses. When the network layer determines the destination station's network address, it forwards the information over a physical network using a MAC address. Different protocol suites use different methods to perform this mapping, the most popular being the Address Resolution Protocol (ARP).

Different protocol suites use different methods for determining the devices MAC address. The three methods most often used are: Address Resolution Protocol (ARP) maps network addresses to MAC addresses, The Hello protocol enables network devices to learn the MAC addresses of other network devices, MAC addresses either are embedded in the network layer address or are generated by an algorithm.

The Address Resolution Protocol (ARP) method is used in the TCP/IP suite. When a network device needs to send data to another device on the same network, it knows the source and destination network addresses for the data transfer. It must somehow map the destination address to a MAC address before forwarding the data. First, the sending station will check its ARP table to see if it has already discovered this destination station's MAC address. If it has not, it will send a broadcast on the network with the destination station's IP address contained in the broadcast. Every station on the network receives the broadcast and compares the embedded IP address to its own. Only the station with the matching IP address replies to the sending station with a packet containing the MAC address for the station. The first station then adds this information to its ARP table for future reference and proceeds to transfer the data.

When the destination device lies on a remote network, one beyond a router, the process is the same except that the sending station sends the ARP request for the MAC address of its default gateway. It then forwards the information to that device. The default gateway will then forward the information over whatever networks necessary to deliver the packet to the network on which the destination device resides. The router on the destination device's network then uses ARP to obtain the MAC of the actual destination device and delivers the packet.

The Hello protocol is a network layer protocol that enables network devices to identify one another and indicate that they are still functional. When a new end system powers up, for example, it broadcasts hello messages onto the network. Devices on the network then return hello replies, and hello messages are also sent at specific intervals to indicate that they are still functional. Network devices can learn the MAC addresses of other devices by examining Hello protocol packets.

Three protocols use predictable MAC addresses. In these protocol suites, MAC addresses are predictable because the network layer either embeds the MAC address in the network layer address or uses an algorithm to determine the MAC address. The three protocols are Xerox Network Systems (XNS), Novell Internetwork Packet Exchange (IPX), and DECnet Phase IV.

8.4.4 Network Layer Addresses

A *network layer address* identifies an entity at the network layer of the OSI layers. Network addresses usually exist within a hierarchical address space and sometimes are called *virtual* or *logical addresses.*

The relationship between a network address and a device is logical and unfixed; it typically is based either on physical network characteristics (the device is on a particular network segment) or on groupings that have no physical basis (the device is part of an AppleTalk zone). End systems require one network layer address for each network layer protocol that they support. (This assumes that the device has only one physical network connection.) Routers and other internetworking devices require one network layer address per physical network connection for each network layer protocol supported. For example, a router with three interfaces each running AppleTalk, TCP/IP, and OSI must have three network layer addresses for each interface. The router therefore has nine network layer addresses.

8.4.5 Hierarchical Versus Flat Address Space

Internetwork address space typically takes one of two forms: hierarchical address space or flat address space. A *hierarchical address space* is organized into numerous subgroups, each successively narrowing an address until it points to a single device (in a manner similar to street addresses). A *flat address space* is organized into a single group (in a manner similar to US Social Security numbers).

Hierarchical addressing offers certain advantages over flat-addressing schemes. Address sorting and recall is simplified using comparison operations. For example, 'Ireland' in a street address eliminates any other country as a possible location.

8.4.6 Address Assignments

Addresses are assigned to devices as one of two types: static and dynamic. *A network administrator according to a preconceived internetwork addressing plan assigns static addresses.* A static address does not change until the network administrator manually changes it. *Devices obtain dynamic addresses* when they attach to a network, by means of some protocol-specific process. A device using a dynamic address often has a different address each time that it connects to the network. Some networks use a server to assign addresses. Server-assigned addresses are recycled for reuse as devices disconnect.

A device is therefore likely to have a different address each time that it connects to the network.

8.4.7 Addresses Versus Names

Internetwork devices usually have both a name and an address associated with them. Internetwork names typically are location-independent and remain associated with a device wherever that device moves (for example, from one building to another). Internetwork addresses usually are location-dependent and change when a device is moved (although MAC addresses are an exception to this rule). As with network addresses being mapped to MAC addresses, names are usually mapped to network addresses through some protocol. The Internet uses Domain Name System (DNS) to map the name of a device to its IP address. For example, it's easier for you to remember www.cisco.com instead of some IP address. Therefore, you type www.cisco.com into your browser when you want to access Cisco's web site. Your computer performs a DNS lookup of the IP address for Cisco's web server and then communicates with it using the network address.

8.5 Introduction to Wireless Data Networks

8.5.1 802.11 Types – What do they all mean?

The IEEE 802.11 standards for wireless LANs have gone a long way toward standardizing wireless LAN development and ensuring a certain level of interoperability, which is absolutely critical for enterprise adoption of WLANs. WLANs' increasing popularity means higher demands on speed, capacity, quality of service, and security. These requirements have resulted in an alphabet soup of 802.11 standards.

8.5.1.1 802.11b

802.11b is the most popular wireless LAN standard. It operates in the unlicensed 2.4-GHz band and provides up to 11 Mbps of data. The data rate is dynamically negotiable and varies depending on various factors, including range (1–2 Mbps up to 400 feet and max of 11 Mbps up to 150 feet).

802.11b's success is partially attributable to the Wi-Fi certification provided by the Wireless Ethernet Capability Alliance (WECA). This certification means that network interface cards (NICs) from any vendor can work with access points from any other vendor. Many companies offer propriety enhancements of 802.11b. Be aware of these because these features may not work with Wi-Fi products from other vendors.

8.5.1.2 802.11a

802.11a is a high-speed WLAN standard that provides speeds of up to 54 Mbps in the relatively uncrowded and unlicensed 5-GHz band. However, actual maximum data rate should be around 22–26 Mbps.

802.11a was recently ratified, and commercial products are now available. A 5-GHz extension of Wi-Fi, called Wi-Fi5, should provide interoperability certification for 802.11a products.

802.11a products are not interoperable with 802.11b products, but since they operate in different bands, they can coexist with a bridging solution. The effective range of 802.11a is lower than 802.11b, but it does have an advantage over 802.11b in terms of speed and capacity.

8.5.1.3 802.11g

802.11g is the high-speed extension of 802.11b in the 2.4-GHz band. Its beauty is its backward compatibility with 802.11b: 802.11g devices can interoperate with existing 802.11b devices.

Technically, 802.11g can operate up to 54 Mbps, but practically, a maximum of around 24 Mbps should be expected. The capacity is, however, still similar to 802.11b with a maximum of three channels operating in parallel.

802.11g is not yet standardized but is expected to become available by the end of 2002.

8.5.1.4 802.11e

802.11e, expected to be approved in the next few months, introduces Quality of Service (QoS) enhancements, which would enable effective voice over IP (a.k.a., voice over LAN) services and streaming multimedia services over 802.11 standards a, b, and g.

8.5.1.5 802.11h

802.11h is expected to become available by the end of 2002 and provides enhancements on top of 802.11a that improve coexistence with other 5-GHz standards like HyperLAN2. This is particularly significant in the European Union, which has been aggressively promoting the HyperLAN2 standard for WLANs.

8.5.1.6 802.11i

802.11i deals with security enhancements, which have suddenly become very significant as the popularity of 802.11b WLANs continues to grow and as several security holes have been identified in WEP, the security element of 8x02.11.

The 802.11i standard should be ratified sometime this year. The final aim of 802.11i is to replace WEP with a new standard called Temporal Key Integrity Protocol (TKIP). In the meantime, we'll see some incremental security upgrades on the existing WEP.

8.6 MAC

802.11, as with other similar networking standards, is comprised of architecture layers. The Media Access Control (MAC) Layer in 802.11 specifies the rules of access to a shared medium. The MAC layer is composed of functional blocks that include mechanisms that provide for contention and contention-free access control on several layers. In order to have Ethernet's true Collision Sense Multiple Access/Collision Detect (CSMA/CD) mechanism would require a full duplex radio that would be too complex and expensive. This solution would still be unable to recognize collisions at the receiving end. As its main contention-free control, the 802.11 uses Collision Sense Multiple Access/Collision Avoidance (CSMA/CA). CSMA/CA is based on a 'listen before talk scheme'.

The way this works is that a station wishing to transmit must first listen to the radio channel to determine if another station is transmitting. If nothing is detected then the transmission proceeds. The CSMA/CA scheme implements a minimum time gap between frames from a given user. Once a frame has been sent from a given transmitting station, that station must wait until the time gap is up to try to transmit again. Once the time has passed, the station selects a random amount of time (called a back-off interval) to wait before listening again to verify a clear channel on which to transmit. If the channel is still busy, another back-off interval is selected that is less than the first. This process is repeated until the waiting time approaches zero, and the station is allowed to transmit. Stations that are hidden from each other present another problem, in order to overcome that, a secondary

FC	Duration ID	Address 1	Address 2	Address 3	Sequence Number	Data	CRC
Bytes used: 2	2	6	6	6	2		4

Figure 8.3 The 802.11 frame format is in eight sections from Frame Control (FC) to checksum (CRC) with total overhead of 28 bytes

contention mechanism is also included in the standard: Network Allocation Vector (NAV). This uses timed transmission, where the stations, in effect, book a slot of certain duration, and other stations stay clear of the medium for the duration. Short frames of the form Request To Send/Clear to Send (RTS/CTS) are used so as not to block the medium for too long.

For an overview of the 802.11 frame format refer to Figure 8.3. The 802.11 frame format is in eight sections with the duration ID built in for use with NAV operation. Beacon frames are sent by the access point to provide information for new stations as well as to keep synchronization. The timed spacing between frames is specified as Short Inter Frame Space (SIFS) or DIFS, that is, Distributed coordination function Inter Frame Space.

Roaming, load balancing and the Simple Network Management Protocol (SNMP) are not incorporated into the 802.11 standard but many vendors are offering some or all of these. It should also be noted that interoperability between various manufacturers' access points have not been addressed in 802.11, but tests have been done by groups of vendors for DHSS and FHSS WLANs respectively. (The two types cannot interoperate.)

8.7 PHY

The bottom of the OSI stack is where we find the PHY. The PHY is the interface between the MAC and wireless media that provides the transmission and reception of data frames over a shared wireless media. There are three levels of functionality provided by the PHY. These are:

1. The PHY layer provides a frame exchange between the MAC and PHY under the control of the physical layer convergence procedure (PLCP) sublayer.
2. The PHY uses signal carrier and spread spectrum modulation to transmit data frames over the media under the control of the physical medium dependent (PMD) sublayer.
3. The PHY provides a carrier sense indication back to the MAC to verify activity on the media.

8.7.1 Direct Sequence Spread Spectrum (DSSS) PHY

One of the three PHY layers supported in the 802.11 standard is the DSSS PHY. The DSSS PHY uses the 2.4 GHz frequency band as the RF transmission media. Data transmission over the media is controlled by the DSSS PMD sublayer as directed by the DSSS PLCP sublayer. The DSSS PMD takes the binary bits of information from the PLCP protocol data unit (PPDU) and transforms them into RF signals for the wireless media by using carrier modulation and DSSS techniques.

8.7.2 The Frequency Hopping Spread Spectrum (FHSS) PHY

FHSS PHY is also one of the three PHY layers supported in the 802.11 standard and uses the 2.4 GHz spectrum as the transmission media. Data transmission over the media is controlled by the FHSS PMD sublayer as directed by the FHSS PLCP sublayer. The FHSS PMD takes the binary bits of information from the whitened PSDU and transforms them into RF signals for the wireless media by using carrier modulation and FHSS techniques.

8.7.3 Infrared (IR) PHY

Another of the three PHY layers supported in the standard is the IR PHY. The IR PHY differs from DSSS and FHSS because IR uses near-visible light as the transmission media. IR communication relies on light energy, which is reflected off objects or by line-of-sight. The IR PHY operation is restricted to indoor environments and cannot pass through walls, such as DSSS and FHSS radio signals. Data transmission over the media is controlled by the IR PMD sublayer as directed by the IR PLCP sublayer. The IR PMD takes the binary bits of information from the PSDU and transforms them into light energy emissions for the wireless media by using carrier modulation.

8.7.4 Physical Layer Extensions to IEEE 802.11

In the Fall of 1997 the IEEE 802 Executive Committee approved two projects that provide for higher rate physical layer (PHY) extensions to IEEE 802.11. Both PHY are defined to operate with the existing MAC.

1. IEEE 802.11a, defines requirements for a PHY operating in the 5.0 GHz U-NII frequency and data rates ranging from 6 Mbps to 54 Mbps.

2. IEEE 802.11b, defines a set of PHY specifications operating in the 2.4 GHz ISM frequency band up to 11 Mbps.

8.7.5 Geographic Regulatory Bodies

WLAN IEEE 802.11-compliant DSSS and FHSS radios that operate in the 2.4 GHz frequency band are required to comply with the local geographical regulatory domains before operating in this spectrum.

WLAN IEEE 802.11-compliant DSSS and FHSS radios are also subject to certification. The technical requirements in the IEEE 802.11 standards have been developed to comply with the regulatory agencies in North America, Europe, and Japan. The regulatory agencies in these regions set emission requirements for WLANs in order to minimize the amount of interference a radio can generate or receive from another in the same type of IEEE 802.11-compliant products. Product developers have the responsibility to check with the regulatory agencies and in some cases there are additional certifications that are necessary for regions within Europe or outside of Japan or North America.

Listed below are some agencies defined by IEEE 802.11.

North America
  Approval Standards: Industry Canada
  Documents: GL36
  Approval Authorities: Federal Communications
  Commission, (FCC)
  USA Documents: CFR 47, Part 15 Sections 15.205,
  15.209, 15.247
  Approval Authority: Industry Canada, FCC (USA)
Spain
  Approval Standards: Supplemento Del Numero 164 Del
  Boletin Oficial
  Del Estado (Published 10, July 91, Revised 25 June 93)
  Documents: ETS 300–328, ETS 300–339
  Approval Authority: Cuadro Nacional De Atribucion
  De Frecuesias
Europe
  Approval Standards: European Telecommunications
  Standards Institute
  Documents: ETS 300–328, ETS 300–339
  Approval Authority: National Type Approval
  Authorities

8.8 The 802.11 Standards (WLAN or WI-FI)

WLANs are typically wireless extensions of wireline LANs. Like its wired counterpart, IEEE P802.3 LAN (Ethernet), a WLAN's primary objective is to provide the type of infrastructure data services typically available through a LAN to the client devices. Once a client device joins a WLAN, it typically stays connected to it until the client device moves away from its boundaries. These devices typically operate in an office or an industrial warehouse, a home, or similar infrastructure.

WLANs do not have an inherent or implied lifespan. They have 'existence' independent of their constituent devices. If all the devices migrated out of a WLAN's coverage area and replacement units arrived, the WLAN would be said to have uninterrupted existence. This is not true for 802.15 WPANs. If the Master does not participate, the network no longer exists.

8.8.1 Defining Wireless LAN Requirements

When deploying wireless LANs it is very important to define the requirements at the beginning of the project. If you don't do this, you'll likely install a solution that doesn't fully meet the needs of users or effectively interface with other systems. You need to get this right the first time because it's difficult and costly to re-engineer the solution once it's already in place.

Requirements define what the wireless LAN must do, which offers the foundation for the design. It's best to leave technical decisions, such as whether to use 802.11b or 802.11a to the design phase, which defines how the solution will work, what components are necessary, etc.

Not sure what constitutes requirements? Let's take a closer look at some common wireless LAN requirements in the order in which you should define them.

(1) **Facility.** Provide a facility description that includes the floor plan, type construction, and possible locations for mounting access points. Find or create building drawings and walk through the facility to verify accuracy. Also, consider taking photos if the building has multiple floors or has a complex layout, such as a five story multi-wing hospital. In addition to a visible inspection, consider performing an RF site survey to complete the facility assessment. All of this will capture the environment in a way that will help you choose the right design alternatives.

(2) **Applications.** Ultimately, the wireless LAN must support user applications, so be sure to fully define them in the requirements. This could be general office applications, such as web browsing, e-mail, and file transfer.

Or it could be wireless patient monitoring in a hospital or price marking in a retail store. Be as specific as possible by defining information types (i.e. data, video, voice) and how they will flow throughout the facility. Application requirements enable you to specify throughput and data rates when designing the system.

(3) **Users.** Don't forget to identify the number of users and where they will use the wireless LAN. Be sure to identify whether users are mobile or stationary, which provides a basis for including roaming in the design. Mobile users will move about the facility and possibly roam across IP domains, creating a need to manage IP addresses dynamically. Some users, however, may be stationary, such as wireless desktops.

(4) **Coverage areas.** This describes where users will need access to the wireless LAN. They might only need connectivity in their offices and conferences rooms, but they might be able to do without wireless connectivity inside power utility rooms and the cafeteria. By properly specifying coverage area, you'll avoid the unnecessary expense of installing access points where they're not needed. Unless obvious, also identify the country in which the wireless LAN will operate. This impacts channel planning and product availability.

(5) **Security.** Describe the sensitivity of the information being stored and sent over the wireless network. You might need to identify a need for encryption if users will be transmitting sensitive information, such as credit card numbers, over the wireless LAN. Be certain to include protection from 'war drivers' who can eavesdrop on your laptop throughout a wireless LAN by including requirements for personal firewalls. Give security requirements plenty of thought so that you design a solution that will protect the company's valuable information.

(6) **End user devices.** You should specify the end user devices (e.g. hardware and operating system) to ensure the solution accommodates them. For example, you could specify that users will have laptops running WindowsXP operating system or a particular brand of PocketPCs having WindowsCE with CompactFlash interfaces. This provides a basis for deciding on the type of 802.11 NIC and drivers to use, as well as assessing the type of middleware that you can use.

(7) **Battery longevity.** An 802.11 NIC will draw current at a couple of hundred milliamps. Batteries under this load will last from a couple of hours to a day or so, depending on the size of the battery. These are constraints for most applications, but it is beneficial to indicate the amount of battery life that users will realistically need. In the design, you can utilize this information to decide whether to activate power management, specify larger batteries, or determine an effective battery-charging plan.

(8) **System interfaces.** In most cases, users will need to access information located in servers on the wired-side of the system. As a result, describe applicable end-systems and interfaces so that you can properly design the wireless system interfaces. For example, you may find that users will need to interface with a warehouse management system on an IBM AS/400. This will later prompt you in the design to consider interface alternatives, such as 5250 terminal emulation and middleware connectivity for interfacing with the AS/400.

(9) **Funding.** The requirements stage of a wireless LAN project is a good time to ask how much money is available. If funding limits are known, then you'll know how much you have to work with when designing the system. In most cases, however, a company will ask how much the system will cost. You'll then need to first define the requirements and design the system before giving a cost estimate.

(10) **Schedules.** Of course a company will generally want the wireless LAN installed 'yesterday,' but we all know that this is an impossibility. You'll need to nail down a realistic completion date, though, and plan accordingly. For example, you may be defining your requirements in July, and a retail store will likely demand that a wireless price marking application be installed by the end of September. This makes it possible to make use of the system during the Christmas holiday season.

After defining these elements, you should have enough information to design the solution. Before proceeding, though, ensure you have consensus from all stakeholders, such as executives, users, and operational support. If requirements are not clear enough, you might consider doing some prototyping or pilot testing to fully understand requirements before spending a lot of money on the design and installation.

Also consider 'baselining' the requirements by getting proper sign-offs and adopt change control procedures. You should be somewhat open to changes, but keep everything under control so that the scope of the project doesn't mushroom into something you can't afford or don't even need!

8.8.2 Minimizing 802.11 Interference Issues

8.8.2.1 What is the Impact of RF Interference?

As a basis for understanding the impact of RF interference in wireless LANs, let us quickly review how 802.11 stations (radio cards and access points) access the medium: Each 802.11 station only transmits packets when there is no other station transmitting. If another station happens to be sending a packet, the other stations will wait until the medium is free. The actual protocol is somewhat more complex, but this gives you enough of the basic concepts.

RF interference involves the presence of unwanted, interfering RF signals that disrupt normal system operations. Because of the 802.11 medium access protocol, an interfering RF signal of sufficient amplitude and frequency can appear as a bogus 802.11 station transmitting a packet. This causes legitimate 802.11 stations to wait for indefinite periods of time until the interfering signal goes away.

To make matters worse, an interfering signal generally does not abide by the 802.11 protocols, so the interfering signal may start abruptly while a legitimate 802.11 station is in the process of transmitting a packet. If this occurs, the destination will receive the packet with errors and not reply to the source station with an acknowledgment. In return, the source station will attempt retransmitting the packet, adding overhead on the network.

Of course this all leads to delays and unhappy users. In some causes, 802.11 will attempt to continue operation in the presence of RF interference by automatically switching to a lower data rate, which slows the use of wireless applications. The worst case, which is fairly uncommon, is that the 802.11 stations will hold off until the interfering signal goes completely away, which could be minutes, hours, or days.

8.8.2.2 Sources of RF Interference that may cause Problems

For 2.4 GHz wireless LANs, there are several sources of interfering signals, including microwave ovens, wireless phones, Bluetooth enabled devices, and other wireless LANs. The most damaging of these are 2.4 GHz wireless phones that people are starting to use in homes and some companies. If one of these phones is in use within the same room as an 802.11b wireless LAN, then expect poor wireless LAN performance.

Microwave ovens operating within 10 feet or so of an access point or radio-equipped user will generally just cause 802.11b performance to drop. Bluetooth-enabled devices, such as laptops and PDAs, will also cause performance degradations if operating in close proximately to 802.11 stations, especially if the 802.11 station is relatively far (i.e. low signal levels) from the station that it's communicating with. The 802.11 and 802.15 standards groups, however, are working on a standard that will enable the coexistence of Bluetooth and 802.11 devices. Other wireless LANs, such as one that your neighbor may be operating, can cause interference unless you coordinate the selection of 802.11b channels.

8.8.2.3 What can be done about RF Interference?

(1) **Analyze the potential for RF interference.** Do this before installing the wireless LAN by performing an RF site survey using tools we've discussed in a previous article. Also, talk to people within the facility and learn about other RF devices that might be in use.

(2) **Prevent the interfering sources from operating.** Once you know the potential sources of RF interference, you could eliminate them by simply turning them off. This is the best way to counter RF interference; however, it's not always practical. For example, you can't tell the company in the office space next to you to shut off their wireless LAN; however, you might be able to disallow the use of Bluetooth-enabled devices or microwave ovens where your 802.11 users reside.

(3) **Provide adequate wireless LAN coverage.** One of the best remedies for 802.11b RF interference is to ensure the wireless LAN has strong signals throughout the areas where users will reside. If wireless LAN signals get too weak, then interfering signals will be more troublesome. Of course this means doing a thorough RF site survey to determine the most effective number and placement of access point.

(4) **Set configuration parameters properly.** If you're deploying 802.11b networks, then tune access points to channels that avoid the frequencies of interfering signals. This might not always work, but it's worth a try. For 802.11 frequency hopping systems, try different hopping patterns. By the way, the newer 802.11e MAC layer, slated for availability sometime in 2002, offers some built-in RF interference avoidance algorithms.

(5) **Deploy the newer 802.11a wireless LANs.** Most potential for RF interference today is in the 2.4 GHz band (i.e. 802.11b). If you find that other interference avoidance techniques don't work well enough, then consider deploying 802.11a networks. At least for the foreseeable future, you can avoid significant RF interference in 802.11a's 5 GHz band. You'll also receive much higher throughput; however, the limited range requires additional access points and higher costs.

8.8.3 Multipath Propagation Defined

Multipath propagation occurs when an RF signal takes different paths when propagating from a source (e.g. a radio NIC) to a destination node (e.g. access point). While the signal is en route, walls, chairs, desks, and other items get in the way and cause the signal to bounce in different directions. A portion of the signal may go directly to the destination, and another part may bounce from a chair to the ceiling, and then to the destination. As a result, some of the signal will encounter delay and travel longer paths to the receiver.

Multipath delay causes the information symbols represented in an 802.11 signal to overlap, which confuses the receiver. This is often referred to as intersymbol interference (ISI). Because the shape of the signal conveys the information being transmitted, the receiver will make mistakes when demodulating the signal's information. If the delays are great enough, bit errors in the packet will occur. The receiver won't be able to distinguish the symbols and interpret the corresponding bits correctly.

When multipath strikes in this way, the receiving station will detect the errors through 802.11's error checking process. The CRC (cyclic redundancy check) checksum will not compute correctly, indicating that there are errors in the packet. In response to bit errors, the receiving station will not send an 802.11 acknowledgment to the source. The source will then eventually retransmit the signal after regaining access to the medium.

Because of retransmissions, users will encounter lower throughput when multipath is significant. The reduction in throughput depends on the environment. As examples, 802.11 signals in homes and offices may encounter 50 nanoseconds multipath delay while a manufacturing plant could be as high as 300 nanoseconds. Based on these values, multipath isn't too much of a problem in homes and offices. Metal machinery and racks in a plant, however, provide a lot of reflective surfaces for RF signals to bounce from and take erratic paths. As a result, be wary of multipath problems in warehouses, processing plants, and other areas full of irregular, metal obstacles.

8.8.4 A Typical Design and Deployment

8.8.4.1 Overview

This document outlines the network design and the management capability necessary to deploy a wireless broadband network in facility partners around the country for Internet access. This design details the physical specifics of the facility partner locations. The heart of the solution is the ORiNOCO Access Server®, which utilizes the 802.11b IEEE standard for *wireless* infrastructures in the unlicensed 2.4 GHz spectrum. The ORiNOCO Access Server eliminates the need for Ethernet wire to the end-user, eliminating the high cost of fully extending Category 5 cabling to each end device. The ORiNOCO Access Server client card is the same as the Lucent WaveLAN client card that has been enhanced with software. The public access version allows for secure access that can be restricted to paying subscribers without providing access to the public at large. Each paying subscriber authenticates with Remote Authentication Dial-In User Server (RADIUS) upon entering a coverage area within a participating facility partner.

8.8.4.2 Network Topology

The (ORiNOCO Wireless Client) network will be built with a single purpose in mind – to provide Internet access to subscribers. Any services available through a traditional dial-up Internet connection are available to the authorized subscriber. This includes access to the subscribers corporate network ('intranet') provided it is currently accessible over the Internet using VPN technologies such as PPTP, SSL, or ssh.

The basis of the network at the facility partner locations is straightforward. It consists of a router, Ethernet switches, and wireless access servers. The elements that add complexity to the design are the variations in architecture and building material of each of the facilities to be covered.

8.8.4.3 Facility Partners

The facility partners consist of several different groups that include hotels, airports, conference centers, and multi-dwelling units.

8.8.4.4 Facility Equipment

The equipment utilized at each facility partner is outlined in Figure 8.4.

8.8.4.5 Power

A single 110VAC 20 amp outlet is required in the Main Distribution Frame (MDF) and Intermediate Distribution Frame (IDF) closets.

8.8.4.6 RF Design

Due to the specific nature of Radio Frequency (RF) communication, the kind of physical environment in which ORiNOCO Access Server will be installed is important. In buildings we generally can distinguish three types of environments:

Components and Part Numbers	
Product	**Part Number**
Lucent Access Point 450 router	AP-ET1-8010
Cajun P333T stackable switch (24 ports)	108563123
Cajun P120 stackable switch (12 ports)	
Orinoco AS-1/2000 wireless access server	407-0031794M
Orinoco WaveLAN Card (Silver)	PC24E-H-FC
Orinoco External Antenna	
PowerDsine power hub (12 ports)	
PowerDsine 48/5VDC Step Down Converter	
Liebert PowerSure 700 Rack Mount UPS	PS70RM-120
Liebert PowerSure Proactive 350 UPS	PSA350-120
Miscellaneous Systemax Cabling and Racking Equipment	

Figure 8.4   Conceptual design components and available part numbers

(1) *Open*

An open environment does not have any area on the path that an RF signal should cover.

(2) *Semi-Open*

A semi-open environment is an area that has partitions that can block RF signals.

(3) *Closed*

A closed environment is an area that has floor-to-ceiling walls or other obstructions in the building such as elevator.

In both the semi-open and closed environments, the actual achievable point-to-point distances largely depend on the construction materials of the obstructing walls and/or partitions.

The less RF barriers present in the environment, the higher the chances are that the performance will be satisfactory throughout the building. RF barriers could be used to separate two ORiNOCO Access Server segments giving both LAN segments maximum performance capacity (Figure 8.5).

RF Barrier description	RF Barrier severity	Examples
Air	Minimal	
Wood	Low	Partitions
Plaster	Low	Inner walls
Synthetic material	Low	Partitions
Asbestos	Low	Ceilings
Glass	Low	Windows, booths
Water	Medium	Damp wood, aquarium
Bricks	Medium	Inner and outer walls
Marble	Medium	Inner walls
Paper	High	Paper rolls, e.g. for newspaper printing
Concrete	High	Floors, outer walls
Bullet-proof glass	High	Security booths
Metal	Very high	Desks, metal partitions, re-enforced concrete

Figure 8.5   RF barrier descriptions

Before proceeding to create the network RF plan, the following items should be checked:

- Determine floor-to-ceiling distance and determine if this is less than 35 feet.
- Determine wall-to-wall distance (e.g. in an open environment) and determine if this is less than 165 by 165 feet.
- Determine the number and kind of partitions and walls in the Access server-to-client paths.
- Determine the kind of environment.
- Determine if the distance is less than set for that environment, i.e. determine if the RF path is qualified or not.

Special attention should be paid to obstructing elevator shafts (metal), 'soft' partitions that contain metal constructions and equipment that causes in-band interference like theft protection equipment, microwave ovens (only 2.4 GHz), copiers and elevator motors.

A typical environment is considered to have a ceiling height of less than 35 feet (10 meters) and open space (wall-to-wall) distances up to 165 feet (50 meters).

In such an environment, minimum disturbance can be expected. The following physical environments have been identified as providing excellent ORiNOCO Access Server performance.

8.8.4.7 Open Environment

An environment without partitions between the ORiNOCO Access Server network nodes. In this environment there are no RF barriers to obstruct the radio. This is an excellent indoor or typical outdoors environment.

Reliable link distances: 400 feet/120 meters or better.

8.8.4.8 Semi-Open Environment

An environment with half-height partitions between the ORiNOCO Access Server network nodes. In this environment, the radio waves are partially obstructed by the partitions.

Reliable link distances: 85 feet/25 meters or better.

The actual constructions of the half-height partitions determine the achievable distance. The specified distances reflect partitions that are constructed of materials that absorb only a limited amount of RF signals such as wood and plastic.

Reliable link distances: 85 feet/25 meters or better.

8.8.4.9 Closed Environment

An environment with floor-to-ceiling walls between the ORiNOCO Access Server network nodes.

Reliable link distances: 50 feet/15 meters or better.

The actual constructions of the walls determine the achievable distance. The specified distances are based on walls that are constructed of materials that absorb only limited RF signals like bricks and plaster.

8.8.4.10 Concrete Walls Environment

An environment with concrete walls between the ORiNOCO Access Server nodes.

Examples of concrete walls are poured reinforced concrete or pre-fabricated reinforced concrete walls.

Physical Environment				
Type	Barriers		Reliable distance	Probable distance
Open	None		120(400)	200(600)
Semi-open	Low severity barriers (partitions of wood/ synthetic material)		30(100)	50(160)
Closed	Medium severity barriers (floor-to-ceiling walls of brick and plaster)		15(50)	25(80)
Obstructed	High severity barriers (metal constructions, reinforced concrete walls)		None	10(30)

Figure 8.6   Summary of point-to-point distances

Reliable link distances: 30 feet/10 meters or better.
The specified distances are based on multiple concrete partitions.
Summaries of the reliable link distances are included in Figure 8.6.

8.8.4.11 Determining if a Bridge Extension is Required

Based on the type of environment and location of critical RF barriers, you can determine whether:

- the entire site can be wireless; or
- a bridge is required to provide wireless networking support for clients that are located beyond the specified distances or behind certain obstructions.

8.8.4.12 Entire Site can be Wireless

See Figure 8.7 for an example of a wireless site. The entire site can be wireless for the planned installation when:

- No severe RF barriers in the path between the ORiNOCO Access Server (or ORiNOCO Access Server bridge) and clients.
- Maximum distance from ORiNOCO Access Server (or ORiNOCO Access Server bridge) to clients is less than distance specified for that office environment.

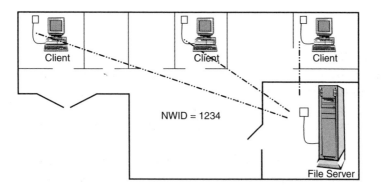

Figure 8.7   Example of a wireless site

If any of the clients fall outside of the distance recommendation or there is a major obstruction in the area between the access server and any or all of the clients, an Access bridge or router will be required to provide wireless networking support for those clients. In case of doubt, ORiNOCO Access Server point-to-point diagnostic measurements should be performed.

8.8.4.13 Space

IDF space

Units are to be Rack Mounted in such a way as to prevent spill damage and provide for physical security.

MDF space

The core switch and WAN router are expected to reside in the same rack or cabinet together in the site's MDF. The racks to be used are made of lightweight aluminum measuring 7 feet high (2.13 m) and 19 inches wide (48.26 cm), the rack offers easy access through front and rear cable segregation. It has double-sided vertical cabling sections for high-density wire management, enhanced by 3-inch-deep (7.62 cm) rack channels. As such, Category 5 cabling will be utilized to connect the 100BaseTX RJ-45 port of the router to the respective 100BaseTX RJ-45 port on the switch. The basement or first Floor (where applicable) houses the MDF Closet. This will be the main access point to the (ORiNOCO Wireless Client) network.

8.8.4.14 Cabling

SYSTIMAX SCS: Category 5 1061 LAN cable supports data, voice, and imaging communications. Cat5 certification will certify cabling up to 100 Mbps. Vendors providing twisted pair and fiber patch cabling must be Systimax certified and a copy of the certification must be provided. The reason for this is that the SYSTIMAX warranty covers defective products, plus the labor cost of fixing the problem, plus a guarantee that SYSTIMAX Structured Connectivity Solution will meet or exceed all EIA/TIA 568-A and ISO/IEC IS 11801 standard. To ensure that the distance of copper is not exceeded, we recommend that the maximum copper distance be limited to 90 meters.

Cables shall be run between floors and through conduits/passages utilizing the path of least resistance while maintaining building codes.

8.8.4.15 Internal Connectivity

Each facility partner site will consist of a Lucent Access Point 450 router, Cajun P333T (core) and Cajun P120 (workgroup) Ethernet switches, and Orinoco AS-1/2000 wireless access servers, as well as the Liebert PowerSure Interactive 700 Rack mount UPS and the Liebert PowerSure Proactive 300 UPS. The specific number of switches will depend on the size of the site itself. At a minimum, there will be one Cajun P333T Ethernet switch that represents the facility backbone (core). The core switch will serve as the connection between the individual wireless access servers and the WAN router. The WAN router will provide the ingress and egress point for backoffice services such as AAA authentication, IP address assignment, network management traffic, and subscriber Internet access.

The core switch and WAN router will reside in the same rack or cabinet in the facilities MDF along with a Liebert PowerSure Interactive 700 Rack mount UPS. As such, Category 5 cabling will be utilized to connect the 100BaseTX RJ-45 port of the router to the respective 100BaseTX RJ-45 port on the switch. The 100BaseTX ports on each network device will be specifically configured (e.g. automatic speed and duplex negotiation disabled) for full duplex to ensure trouble-free operation and an aggregate throughput of 200 Mbits/s. The core switch in the MDF will feed each individual IDF closet.

The individual wireless access servers will be distributed throughout site location to achieve the best possible signal strength and continuous cell coverage. (This will be determined by the individual site survey results for each facility partner location.) Given the use of Power Over Ethernet (POEt), traffic from the wireless access servers will first be carried to a PowerDsine Power Hub before being terminated on the core switch where it will be either routed to the back office, within the facility itself, or to the Internet.

The physical means of making this connection will depend on the distance of the access server from the MDF closet. For access servers within the Ethernet distance limitations of copper, a direct connection to the PowerHub with Category 5 cabling will be used.

Access servers outside the Ethernet distance limitation (100 meters) of copper will use multimode fiber from the core switch to a workgroup switch residing in an IDF. The transition from fiber to copper will require a pair of media converters. To make the connection to the core switch in MDF, a single 10BaseTX RJ-45 port from the workgroup switch will be connected to the media converter for the transition from copper to fiber. The fiber jumper from the media converter will be connected to the appropriate fiber pair on the FDP and carried by multimode fiber through the inter-floor riser conduit to another FDP in the MDF. Once in the MDF a fiber jumper will connect the appropriate fiber pair to another rack mounted media converter for the transition from fiber back to copper. The respective 10BaseTX RJ-45 port on the media converter will then be connected to the appropriate 10BaseTX RJ-45 port on the core switch for that IDF with Category 5 cabling. Once the transition to copper has occurred the cable run will connect to an RJ-45 port on the Cajun P120 workgroup switch.

With the use of Power Over Ethernet (POEt), the RJ-45 ports on the Cajun P120 switch that ordinarily connect directly to the access servers will instead connect to the 'data only' ports on the PowerDsine Power Hub. The PowerDsine Power Hub will add 48VDC to the spare pairs 4,5,7 and 8 and then exit the Power Hub through the corresponding 'data and power' ports on PowerDsine Power Hub. The cable attached to the 'data and power' port will pass through a converter that will step down the applied voltage to 5VDC before attaching to the wireless access server.

If more than one Access Server is being fed from an individual IDF closet, they will share an access server and Cajun P120 in the IDF.

8.8.4.16 External Connectivity

Each facility partner location will be connected to the Internet service provider network by an individual T1 point-to-point circuit operating at 1.534 Mbits/s. The serial interface of the Lucent Access Point 450 router, with its internal CSU/DSU will terminate this circuit at the facility partner. The interface will be configured for the Point-to-Point Protocol (PPP) encapsulation type to ensure interoperability with the router of the chosen Internet service provider.

8.8.4.17 Traffic Flow

The traffic within this network consists of two distinct flows: management and subscriber traffic.

The management traffic consists of elements such as AAA, SNMP, telnet and tftp for configuration and maintenance. Since the facility partner and back-office will not be directly connected – each will be independently connected to the Internet – the plan is to establish a VPN tunnel between facility partner router and the back-office router. This ensures that management traffic that traverses the Internet will do such in a secure manner.

The target audience is defined as business travelers; the traffic is expected to be predominately e-mail and web traffic. Subscriber traffic will be sent directly to the Internet from the facility partner location once back-office functions are completed for the session.

8.8.4.18 Routing

Due to the design and heavy utilization of the Internet for WAN transport, routing will be restricted to static routes from each facility partner location to the upstream service provider. Should the architecture change to include redundancy or the WAN transport change this topic will need to be revisited. Clients will be required to provide their own VPN tunneling capabilities.

8.8.4.19 IP Addressing and Assignment

Given the time constraints of the project and the justification necessary to acquire address space from the American Registry for Internet Numbers (ARIN) it will be necessary for (ORiNOCO Wireless Client) to acquire and utilize public address space from their service provider. Based on the initial take rate at each facility partner, one or more address blocks equal to 64 (62 usable) addresses (/26) will be required.

The assignment of IP addresses to subscribers will be made by QIP during AAA authentication process.

8.8.4.20 Security

Due to the nature of (ORiNOCO Wireless Client) service offering ('unencumbered Internet access') security is designed to be relaxed with regard to the individual subscribers. The exception is the use of encryption of the wireless subscriber traffic. Inclusion of security for purposes of protecting the individual subscriber has the potential of affecting the current and future functionality of the subscriber.

The (ORiNOCO Wireless Client) network infrastructure and application servers will be secured using a variety of measures. Outlined below are descriptions of these security measures that will be taken throughout.

8.8.4.21 Subscriber Authentication, Authorization and Accounting

NavisRadius will be used for subscriber Authorization, Authentication and Accounting (AAA) to prevent unauthorized use of the (ORiNOCO Wireless Client) wireless network. All users will be required to authenticate before being allowed entry to the network. Non-subscribers that happen to be in possession of an 802.11b compliant network card will be prevented from using the service by the same authentication process.

Challenge Handshake Authentication Protocol (CHAP) will be deployed. With CHAP, the authenticator sends a randomly generated 'challenge' string to the client, along with its hostname. The client uses the hostname to look up the appropriate secret, combines it with the challenge, and encrypts the string using a one-way hashing function. The result is returned to the server along with the client's hostname. The server now performs the same computation, and acknowledges the client if it arrives at the same result.

Another feature of CHAP is that it challenges at regular intervals to make sure an intruder hasn't replaced the client since the initial challenge. CHAP will be used to ensure the authentication process isn't susceptible to attack.

8.8.4.22 Network Equipment Access

Access lists will be used to restrict access to the routers and switches from the Internet at the facility partner sites. Telnet is the standard method of accessing network equipment remotely. The contents (payload) of the telnet session are sent as clear text. The Lucent Access Point routers support secure shell (ssh) that encrypts the contents (payload) of the session using one of several ciphers. Secure shell will be used to perform any remote diagnostics and configuration of the routers. The Cajun switches being used do not support ssh. The router, using ssh, can serve as an encryption 'gateway' to the products at the facility partner that do not support ssh. Any time a switch is to be accessed remotely for diagnostic and configuration purposes, an ssh session should be started with the router residing on-site and then telnet from the router to the switches as needed. The risk of using telnet between the router and the switch is minimal since the path that this traffic traverses is switched.

Authorized network personnel attempting to access the devices that make up the network infrastructure will be authenticated with AAA. The database containing authorized network personnel will differ from the database containing the (ORiNOCO Wireless Client) subscribers.

8.8.4.23 Physical Security

Outside the Network Operations Center, there are two distinct areas of security that need consideration: the central location and the facility partner sites.

As noted elsewhere in this document the central location will house the back-end servers and network equipment to support them. The central location is slated to reside in a private co-location cage at a service provider neutral data center. This data center will have a variety of security mechanisms (access cards, guards, video surveillance, etc.) to ensure authorized access to the facility. It will be necessary to ensure that the equipment residing within the facility isn't disturbed, either intentionally or accidentally, by utilizing locks on the cabinets and cage.

Security of the equipment placed at the facility partner sites will be more difficult to control. As a result, the primary goal is to prevent casual tampering and theft. Utilizing locking wall cabinets to secure each Access Server installed will accomplish this. In addition, the power, data, and antenna cables feeding the cabinet should be enclosed in metal conduits to prevent service disruption. The core switch and router at the facility partner will be located in a 19″ rack located in a limited access area that will be locked behind a locked door.

The Physical security of the cable plant installation can be enhanced by the use of Fiber cable in lieu of Category 5 copper when needed, as recommended by the US Government. This effectively halts unauthorized tapping, deters casual tampering, and greatly reduces EM radiation. The use of fiber is only needed when it is necessary to install data runs between the equipment closets using mechanical shafts such as elevators or when the maximum copper distance is greater than recommended.

8.8.4.24 Network Redundancy

Virtual Router Redundancy Protocol (VRRP) will be used in the NOC/Back Office to provide redundancy of routing architecture. VRRP provides a way for IP workstations to keep communicating on the Inter/intranet even if their default router becomes unavailable. VRRP works by creating a phantom router that has its own IP and MAC addresses. The workstations use this phantom router as their default router. VRRP routers communicate among themselves to designate one as the active and the other(s) as the standby router. The active router sends periodic hello messages. The other VRRP routers listen for the hello messages, if hello messages are not received, the standby router takes over and becomes the active router. The new active router assumes both the IP and MAC addresses of the phantom and the end nodes see no changes. The end nodes continue to send packets to the phantom router's MAC address and the active router delivers them.

8.8.4.25 Network Management Systems

This paragraph identifies the Network Management Systems required for the interim NOC to support the wireless solution. The NOC processes and staffing plans are developed after the validation tests have been completed.

8.8.4.26 The Open System Interconnect (OSI) Model

Network Management has been defined by the International Standard Organization (ISO) as being comprised of the following five functional areas: Fault, Configuration, Accounting, Performance and Security. This definition or model has been referred to as the FCAPS model.

Using this model as a guide, the Interim NOC will focus on providing functionality in each of these areas. The permanent NOC is expected to have a more robust functionality, particularly in the Performance area. To implement this model, a number of software products are planned for deployment. These software products will be deployed in the Interim NOC. Subsequently, additional copies of this software will be deployed in the permanent NOCs. The difference between software deployed in the interim and permanent NOCs are related to the need for larger capacity licenses in the permanent NOCs and the expected release of newer versions of some of the software products.

The software products to be deployed in the interim NOC are described below. They are organized by the corresponding activity defined in the FCAPS model.

Fault

Hewlett Packard Network Node Manager (NNM). This software product discovers devices supporting IP and SNMP on the network and builds a database of the discovered devices. It builds a graphical, logical map of the network. NNM can be used to test connectivity to the devices in its database and processes trap messages received from the devices. This product also can process unsolicited messages (traps) sent by network infrastructure devices and build a database of the hardware components on the network during the network discovery phase, NNM therefore provides Fault and some Configuration services.

As a main component of the NOC design, Network Node Manager provides a number of capabilities including the following.

- Automatic discovery and monitoring of TCP/IP devices.
- Management of other vendors SNMP devices via MIB objects.
- Collection of historical MIB information about MIB objects for trend reporting.
- Event thresholding for MIB objects.
- Monitors and reports on the status of LAN interface via polling (ICMP echo request or SNMP get).
- Provide an integration platform for Element Managers.

Veritas NerveCenter. This product provides intelligent correlation of traps received from the network. In the case where multiple devices are reporting problems, NerveCenter uses rules previously configured into the software to attempt to determine the root cause of the traps and to determine which traps are related to the same fault. As the brain of the solution, NerveCenter will poll for device status according to predefined rules that help control management traffic. It also will determine a true 'Faulty' device, then generate an alarm and finally kick off the notification process. Following are the functions provide by NerveCenter:

- Event correlation
- Generate alarms
- Provide notification capability
- Submit Trouble Tickets to Remedy ARS
- Provide selective and controlled polling capability.

SiteNet SNMP Manager. This product is used to monitor the status and operating mode of properly equipped Liebert Uninterruptible Power Sources. The software can display the charge status of the batteries and whether the UPS is running on battery power or is in pass-through mode on AC line power. The software integrates with HP NNM.

Remedy's ARS. This product provides many capabilities. In the wireless environment it will be used in conjunction with NerveCenter to generate a Trouble Ticket. It will interact with Visionael to provide details for the Ticket. Remedy's ARS functions are as follows:

- Generate 'intelligent' Trouble Tickets requested by NerveCenter;
- Provide Knowledge Based application foundation.

Visionael, as a physical inventory management platform, is to provide Remedy's ARS trouble Ticket with details as defined by the Helpdesk. It also will provide HP OV-NNM information for graphical tracing, which will be used by helpdesk personnel to verify and confirm the 'Faulty' device. For the network management architecture, Visionael provides the following functions:

- Provide physical inventory management capability;
- Provide ARS with device details;
- Provide HP OpenView with device details for graphical tracing.

Configuration

Lucent CajunView Plus. This product works in conjunction with NNM and it is used to configure and manage software configurations on Cajun Switches. This product will be used in the wireless NOC to manage the Cajun LAN switches at the facility partner locations. The Base package includes modules for a number of Cajun products. The P330 product is not included in the base suite, but a separate module is available for these devices. Other modules in the suite can be selectively installed on the management system.

Lucent AS Manager. This software product provides the ability to configure Access Server 1000 and Access Server 2000s. This product was initially developed for use in an extremely small network, or for use by a support technician working onsite where direct connection to the Access Server is possible. A replacement product is under development. The new product is currently unnamed. Preliminary indications are that the new product will operate in conjunction with HP Network Node Manager and have more enterprise level capabilities.

- The AS Manager will be used to manage the configurations of AS 1000 and AS 2000 access devices installed in the facility partner locations.

Access Point Access View. This product is a JAVA-based management application for Access Point routers, which uses a client server architecture model. The Access Point routers have an embedded web server and the support staff accesses the server using a common web browser, such as Microsoft Internet Explorer or Netscape Navigator.

- Access View provides capability for establishing and maintaining the configuration on the Access Point routers. (Router configurations can also be maintained using a command line interface and the Telnet protocol.)

Liebert SiteNet SNMP Manager. This product is used to monitor the status and condition of Liebert uninterruptible power systems (UPS). The UPS devices will be located in the facility partner location to provide battery backup to the network elements.

Accounting

Lucent NavisRadius. This software performs services in the Accounting and Security activities. In the Accounting activity, this software logs usage by user, which can be transferred to the Keenan system for billing purposes.

Lucent QIP. This software provides a combination of IP address management functions. The first function provided is Dynamic Host Configuration Protocol

(DHCP). This dynamically allocates IP addresses to authenticated users. The allocated IP address and other configured information are sent to the client at login in time. Following the transmission of this data, the second major function is initiated. The second function provided is updating the dynamic Domain Name Service (DNS) with the host name and IP address that was just assigned by DHCP. Lucent QIP is an IP address management system that is designed to provide centralized administration of IP network information, dynamic host configuration protocol (DHCP) address assignment, and dynamic domain name system (DDNS) configuration and management.

Lucent QIP with its many capabilities, will initially provide the network management architecture with the following functions:

- Provide HP NNM with IP address and name resolution;
- Interact with Physical Assets Manager and AAA for IP Address assignments;
- Update DNS servers.

Performance

In the interim NOC, a dedicated performance monitor will not be implemented. In the permanent NOC, VitalSuite, or some portion of the VitalSuite package is being considered.

Security

Lucent NavisRadius. This software authenticates users, which are allowed to use the services provided by ORiNOCO Wireless Client. This software must have information to identify all valid users.

Implementation strategy

The preliminary design for the implementation of the interim NOC is to install the above-listed software products on a series of processors connected to a sub-network of the Wireless network. These processors are a combination of Sun Microsystems servers running Solaris 2.6 and Intel-based servers running Windows NT 4.0. All of the software mentioned above will be installed on the Sun server. The Kenan billing system is expected to be installed on a separate Sun Workstation and is not addressed in this document. The Kenan system will be installed and supported by professional services staff from Kenan.

Redundancy

The computer platforms described herein are single systems and represent single sources of failure. If the Sun server were to fail, the Radius software would be unavailable and users would eventually lose their connection, and additional users would not be able to logon to the network. For this reason alone, a secondary Sun server is required. The second server would be interconnected to the first server and provide the ability to continue processing in the event the first server failed. The systems would be interconnected with a shared hard drive array.

Staffing

In order to support the Interim NOC, the staff must have specific skill sets. These skills support general networking requirements and specific skills and knowledge of the equipment being deployed. The skills are broken into three categories; those required of all staff, those required by field support staff; and those required by the NOC staff.

General skills set

- Familiarity with Routers, Switches, Ethernet, Frame Relay, ATM, T1 and RF technologies.
- Specific knowledge of the ORiNOCO Access Server, Cajun, UPS, and CSU/DSU products.
- Solid troubleshooting procedures.
- Clear and concise communication skills.
- Ability to communicate with client or other designated persons.

Field support staff

- Proper installation procedures for ORiNOCO Access Server, Cajun, UPS, and CSU/DSU products.

- Proper fault isolation and service procedures for the ORiNOCO Access Server, Cajun, UPS, and CSU/DSU products.
- Ability to maintain accurate as built documentation.

NOC staff

- Understanding and ability to use network management platforms.
- Ability to configure and maintain network management platforms.
- Ability to convey fault information to field personnel, and to work with them through fault isolation, repairs, and closure.
- Ability to lead effort to resolve network faults or to resolve other service affecting conditions.

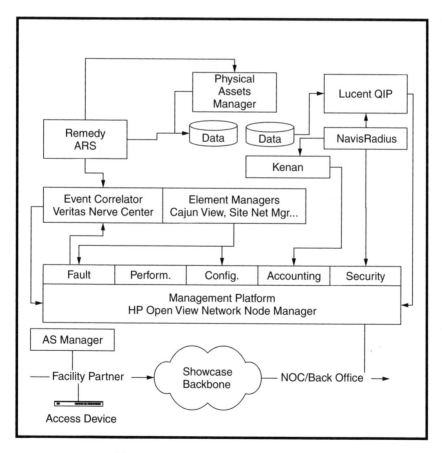

Figure 8.8 Functional model diagram

Training

To enhance the probability of success for a specific engineer, it is likely that the following training will need to be provided.

- Specific installation and maintenance of ORiNOCO Access Server, and other products.

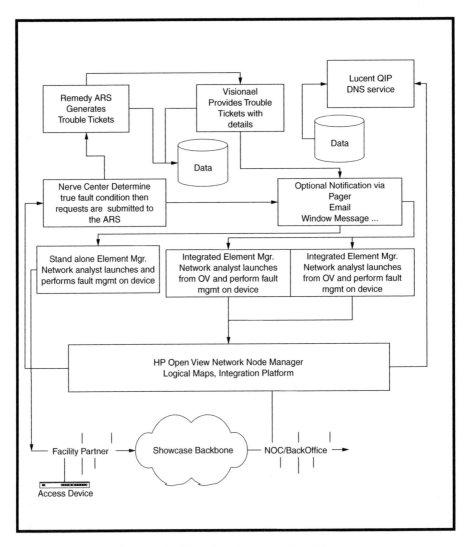

Figure 8.9 Functional process flow diagram

- Specific configuration and maintenance of network management software packages.

8.8.4.27 The Functional Model

This functional model illustrates the four functional areas addressed in the interim phase. The Performance area will be addressed in the design of the permanent NOC.

Functional process flow

This process map gives a high level view of the functional roles of the software to be deployed in the interim NOC.

8.8.4.28 The NOC/Back Office

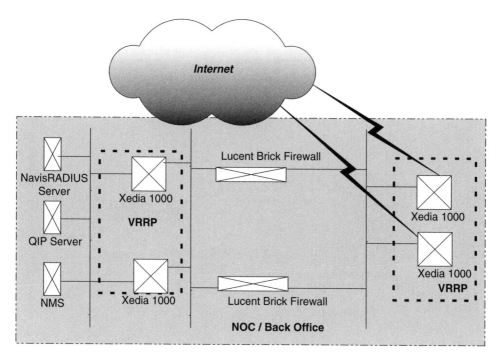

Figure 8.10 The NOC/BackOffice

8.8.4.29 NOC/BackOffice Environment

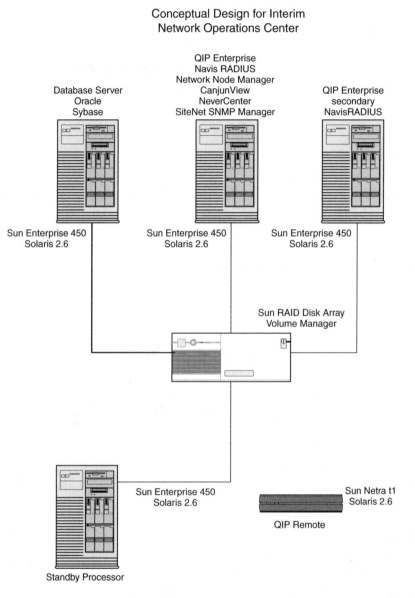

Conceptual Design for Interim
Network Operations Center

Database Server
Oracle
Sybase

QIP Enterprise
Navis RADIUS
Network Node Manager
CanjunView
NeverCenter
SiteNet SNMP Manager

QIP Enterprise
secondary
NavisRADIUS

Sun Enterprise 450
Solaris 2.6

Sun Enterprise 450
Solaris 2.6

Sun Enterprise 450
Solaris 2.6

Sun RAID Disk Array
Volume Manager

Sun Enterprise 450
Solaris 2.6

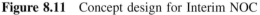

Sun Netra t1
Solaris 2.6

QIP Remote

Standby Processor

Figure 8.11   Concept design for Interim NOC

8.8.4.30 The Network Management Functional Areas

Fault Management   The process of detecting, isolating, and correcting network problems.   Fault management enables network engineers to quickly detect errors and begin recovery from them.
Performance Management   The process of analyzing and controlling data throughput on a network.   Performance management allows network engineers to provide end users with consistent, reliable levels of service.
Configuration Management   The process of obtaining information from a network and setting up devices based on that information.   Configuration management allows centralized control over the configuration of network devices.   From a single console, network engineers can determine and set a variety of hardware and software components for routers, workstations, hups, switches and access devices.
Accounting Management   The process of measuring resource utilization on a network. This data can be used to established metrics, determine costs, bill users and check quotas.   Accounting management allows network engineers to property allocates resources and bill users for their consumption of those resources.
Security Management   The process of controlling access to network resources and sensitve information. This control is effected by limiting access to hosts and network devices, to particular applications on a given device, and to particular protocols that traverse the network.

Figure 8.12 The five functional areas of network management as defined by the International Standard Organization

8.8.4.31 Antenna Location

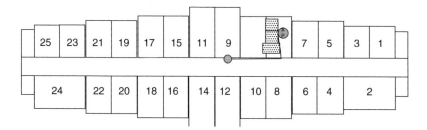

Figure 8.13 Example antenna location – 1

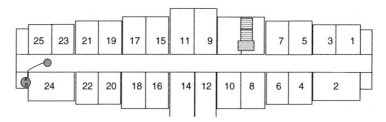

Figure 8.14 Example antenna location – 2

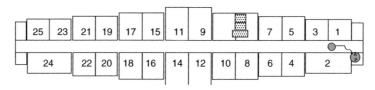

Figure 8.15 Example antenna location – 3

8.8.4.32 Channel Configuration

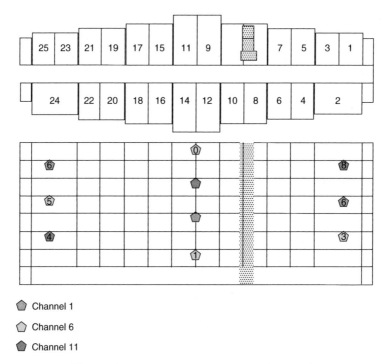

Figure 8.16 Example channel configuration

8.8.4.33 RF Coverage

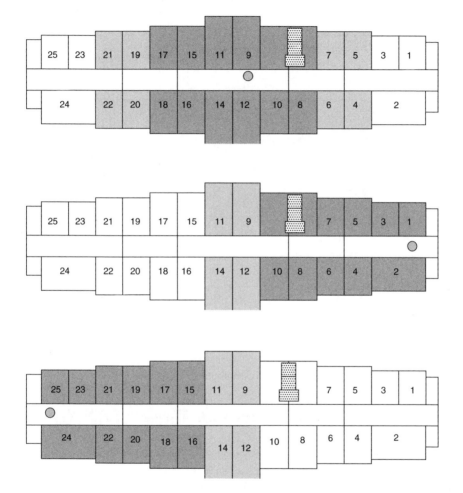

Figure 8.17   Example RF coverage

8.9 Security

In this section, we will look at the security features built into the IEEE 802.11 standards that protect privacy and access control. To enhance 802.11b security, additional security capabilities such as client access control lists, an IP filtering firewall and VPN tunneling are also looked at for implementation.

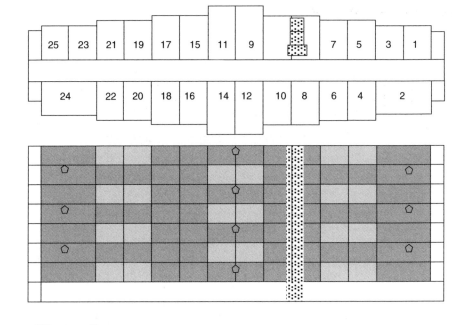

| | SNR 10–15 dB, datarate > = 5.5 Mb |

| | SNR >15 dB, datarate = 11 Mb |

Figure 8.18   Total signal-to-noise ratio

8.9.1 Potential Security Issues with Wireless LAN Systems

We know that, just like any other radio wave, a wireless LAN signal is not limited to the physical confines of a building and that the potential exists for unauthorized access to the network from personnel outside the intended coverage area. Two possible security concerns arise from this aspect of a wireless LAN:

- Unauthorized access to network resources via the wireless media;
- Eavesdropping of the wireless signaling.

8.10 Overview of 802.11b Security Mechanisms

Many hardware vendors have devised proprietary solutions to handle the deficiencies of the 802.11b Standard but they are out of the scope of this document

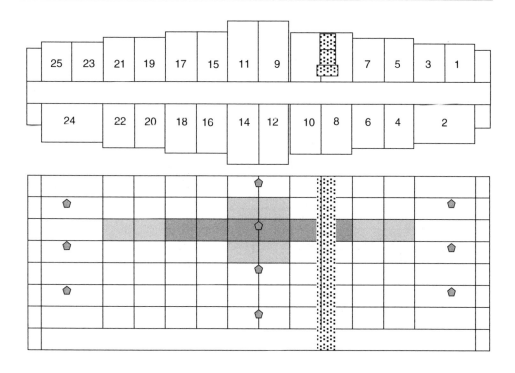

 SNR 10–15 dB,
datarate > = 5.5 Mb

SNR >15 dB,
datarate = 11 Mb

Figure 8.19   Central placement signal-to-noise ratio

and will not be discussed. The 802.11b Standard has two basic security defense
mechanisms. These two mechanisms are discussed next.

8.10.1 SSID – Network Name

A Service Set Identification (SSID) is basically the network name of a Wireless
LAN (WLAN) segment and it is supposed to logically segment the users and

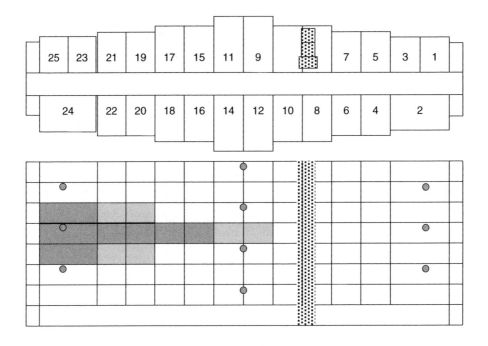

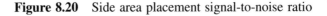

SNR 10–15 dB,
datarate >= 5.5 Mb

SNR > 15 dB,
datarate = 11 Mb

Figure 8.20   Side area placement signal-to-noise ratio

APs. Theoretically, the client's wireless Network Interface Card (NIC) should be
configured with the same SSID as the AP in order to join the network.

8.10.2 WEP – Wired Equivalent Privacy

Wired Equivalent Privacy (WEP) was designed by the IEEE to bring WLAN
security to a level comparable to a wired networking environment such as a
Local Area Network (LAN). WEP uses a security feature widely used throughout
the security industry known as encryption.

WEP's encryption process uses a symmetric key and a mathematical algorithm to convert data into an unreadable format called cipher-text. In cryptography, a symmetric key is a variable length value used to encrypt or decrypt a block of data. Any device needing to participate in the symmetric encryption process must possess the same key. WEP keys are configured by the WLAN administrator and the larger the key, the harder it will be to break the encryption cipher.

RC4 is the encryption algorithm used by WEP and it needs the assistance of an Initialization Vector (IV). An IV is a pseudo-random binary string used to jump-start the encryption process for algorithms that depend on a previous sequence of cipher-text blocks. A smaller IV in conjunction with keys that do not frequently change will increase the chances that encrypted data packets will duplicate the IV.

WEP consists of up to four variable length symmetric keys based on the RC4 stream cipher. All keys are static in nature and are common to all devices on the WLAN. This means that the WEP keys are manually configured on the WLAN devices and will not change until the administrator configures different keys. Most 802.11b equipment comes with two key sizes. The two key sizes are shown below.

- 64-bit 40-bit Key and a 24-bit Initialization Vector;
- 128-bit 104-bit Key and a 24-bit Initialization Vector.

Nonetheless, the static nature of the WEP keys and the small initialization vector combine to create a massive problem in both scalability and security. These are all IEEE standards problems but as stated earlier, many hardware vendors have created proprietary solutions. There are two main purposes of WEP and they can be seen below.

- Deny WLAN Access;
- Prevent Replay Attacks.

An AP will use WEP to prevent WLAN access by sending a text challenge to an end user client. The client is supposed to encrypt the challenge with their WEP key and return it back to the AP. If the results are identical, the user is granted access.

WEP also prevents replay attacks. This is where an attacker will try to decode sniffed data packets. If the intruding WLAN user manages to capture WEP encrypted 802.11b frames out of the air, the attacker will not be able to decode the packets unless they possess the proper WEP key to decrypt the data.

8.11 Authentication and Association

In order for a wireless client to have access to a WLAN, the 802.11b Standard indicates that the client must go through two processes. These two processes are known as:

- the Authentication Process, and
- the Association Process.

Once the wireless client has successfully completed the authentication and association processes, the end user will be given access to the WLAN.

8.11.1 Authentication Process

A wireless client that desires access to a WLAN must first undergo the authentication process. This authentication process validates information about the client and is the initial step in connecting with the wireless AP. The authentication process consists of two types of authentication:

- Open System Authentication;
- Shared Key Authentication.

With Open System Authentication (OSA), all negotiation is done in clear text and it will allow a client to associate with the AP without possessing the proper WEP key. The only thing that is needed is the proper SSID. Some APs will even accept a null SSID. An AP can be configured for OSA but still be configured for WEP data encryption. So if a client properly associates with the AP, the client will be unable to encrypt or decrypt data it receives from the AP.

In contrast to OSA, Shared Key Authentication (SKA) forces the AP to send a challenge text packet to the wireless client. The client in turn will encrypt the challenge text with its WEP key and send it back to the AP. The AP will then decrypt the challenge and compare it to the original text sent. If the two match, the AP will allow the client to associate with it.

8.11.2 Association Process

The Association Process is the course of action in which a wireless client pursues a connection with an AP. The Association Process is the final step in connecting to a wireless AP.

8.11.3 Authenticated and Associated

The 802.11b Standard indicates that the client must first authenticate to the AP and then it must associate to the AP. The standard also specifies that these two aforementioned processes will make up one of three states in the sequence joining a WLAN through an AP. The three states are:

- State 1: Unauthenticated and Unassociated.
- State 2: Authenticated and Unassociated.
- State 3: Authenticated and Associated.

Unauthenticated and unassociated is the initial state of an AP and a client. Once a client has completed the authentication process but has yet to complete the association process, the client is considered to be in the second stage known as authenticated and unassociated. After the client successfully associates with an AP, the client has completed the final state and is considered to be authenticated and associated. The client must be authenticated and associated with an AP before access to a WLAN is granted. There are three phases in the development of a client becoming authenticated and associated with an AP. The three phases that make up this state are:

(1) Probing Phase
(2) Authentication Phase
(3) Association Phase.

8.11.4 Probing Phase

A wireless client will send a probe request packet out on all channels and any AP that is in range of the client will respond with a probe response packet. These AP probe response packets contain information that the client will use in the association process.

8.11.5 Authentication Phase

As stated earlier, the authentication phase can use either OSA or SKA. The configuration of the AP will dictate which type of authentication is used. For the most secure WLAN environment, it is highly recommended to go with SKA authentication.

In the OSA scheme, a client will send an authentication request packet to the AP. The AP will analyze the authentication request packet and send an authentication response packet back to the client stating whether it is allowed to move onto the association phase.

In the SKA scheme, a client goes through the same process as with OSA but the AP sends a challenge text to the client. As stated earlier, the client will take this challenge and use its static WEP key to encrypt the text. Once the client sends it back to the AP, the AP will then decrypt the challenge with its static WEP key and compare it to the original text sent. The AP will allow the client to move on to the association phase if the text was properly decrypted but if the AP found the text to be contradictory, it will prevent the client from accessing the WLAN.

8.11.6 Association Phase

In the association phase, the client will send an association request packet to the AP. The AP will send an association response packet back to the client stating whether the client will be allowed to have access to the WLAN. The 'Authenticated and Associated' state is the final negotiation step between an AP and a wireless client. If there are no other security mechanisms (RADIUS, EAP, or 802.1X) in place, the client will have access to the WLAN.

8.12 Wireless Tools

Wireless LAN installations can be a little tricky. Unlike wired networks, you can't visualize or see the wireless medium. The construction of a facility and silent sources of RF interference impact the propagation of radio waves. This can make it tougher to plan the location of access points.

One of the ways to avoid these drawbacks is to perform an RF site survey using the appropriate site survey tools. These will help you plan access point locations for adequate coverage and resiliency to potential RF interference. There are various types of tools you can use to aid in your endeavor.

8.12.1 Basic Tools

The traditional method for performing an RF site survey includes a laptop equipped with an 802.11 PC Card and site survey software supplied at no additional cost from the radio card vendor. The software features vary greatly by vendor, but a common function among them all displays the strength and quality of the signal emanating from the access point. This helps determine effective operating range (i.e. coverage area) between end users and access points.

This relatively inexpensive site survey tool has some drawbacks. For one, it's physically demanding to lug a laptop around a building all day when doing the testing. You can ease this problem though, by using one of the recently released 802.11 CompactFlash cards along with a pocket PC device, such as the Compaq iPAQ, Casio Cassiopeia, or HP Jornada. This reduces the physical demands of performing the tests, but you'll be lacking a significant capability: the detection of RF interference between access points and from other RF sources, such as Bluetooth devices, microwave ovens, and wireless phones.

8.12.2 Advanced Tools

Advanced 802.11 site survey tools include spectrum analysis which allows you to understand the affects of the environment on the transmission of 802.11 signals. An 802.11b spectrum analyzer graphically illustrates the amplitude of all signals falling within a chosen 22 MHz channel which in turn enables you to distinguish 802.11 signals from other RF sources that may cause interference, making it possible to locate and eliminate the source of interference or use additional access points to resolve the problem.

Another key spectrum analysis feature is the monitoring of channel usage and overlap. 802.11b limits up to three access points to operate in the same general area without interference and corresponding performance impacts, causing difficulties when planning the location and assignment of channels in large networks. Spectrum analysis displays these channels, enabling you to make better decisions on locating and assigning channels to access points.

8.13 Penetration Testing on 802.11

The IEEE 802.11 Standards have left many doors open for hackers to exploit their shortcomings and the goal of this section is to bring light to these issues while looking at how to prevent them.

A technique of attacking wireless networks that hackers have dubbed as 'WarDriving' is becoming an everyday buzzword in the security industry. This is the wireless brother of 'WarDialing' that is done on wired networks. This section will cover the fundamentals on how to deter a WarDriving attack by performing controlled penetration tests on a wireless network.

There is not a lot to do to prepare for penetrating a WLAN. We also try to maintain uniformity in how we conduct penetration testing in the equipment and software used. This allows for ease of duplication among our peers. All network sniffing and penetration testing discussed in this section has been conducted with the following hardware set up:

- Dell Latitude CPH 850 MHz Laptop with 256 MB RAM.
- Microsoft Windows XP Professional Operating System.
- Lucent Technologies WI-FI Orinoco Gold 11 Mbps NIC.

In order to conduct a penetration test on a WLAN, all necessary materials must be collected, installed and configured. Preparing for a wireless penetration testing consists of two steps, which are installing the Orinoco Gold NIC and setting up the Wireless 802.11b Sniffers.

8.13.1 Installing the ORiNOCO NIC

Installing the wireless NIC is a particularly important stage. A wireless NIC that is not correctly installed and configured will not be capable of taking advantage of all WarDriving tricks documented throughout the body of this report. A properly installed Orinoco Gold NIC has two major features that a normal Orinoco Gold NIC doesn't. These two features are:

- Promiscuous network sniffing;
- Ability to change the MAC address.

The NIC should be inserted into the laptop's PCMCIA slot and Windows XP will install its own drivers for the adapter. As a best practice, the PC should be rebooted after installing each driver. The default drivers that Windows XP installs are inadequate for the purposes of WarDriving and need to be hacked with special versions of software and firmware. This process must be carried out in a precise sequence.

First, an older version of drivers and firmware (R6.4winter2001) must be installed from the OrinocoWireless.com or WaveLan.com FTP sites. This is what will allow the NIC to have its Media Access Control (MAC) address manually configured to a custom setting. The drivers will update the firmware and software to:

- Orinoco Station Functions firmware Variant 1, Version 6.16.
- NDIS 5 Miniport driver Variant 1, Version 6.28.
- Orinoco Client Manager Variant 1, Version 1.58.

Once the firmware and software have been updated, a final patch can be applied to the Orinoco NIC. A WildPackets AiroPeek driver is a hacked version of the Orinoco Gold NIC driver that will allows the NIC to sniff promiscuously. Once this driver is properly loaded, the NIC is fully operational for WarDriving.

8.13.2 Setting up the Sniffers

There are several 802.11b Sniffers that can sniff 802.11b frames out of the air. This document only addresses free solutions, as opposed to expensive commercial products. The two sniffers used in this exercise are WinDump and Ethereal.

WinDump and Ethereal were originally UNIX utilities that relied on libpcap, but they have been ported to Win32. In order for the Win32 ports to work, WinPCap must be loaded before the sniffers can pick up traffic. WinPCap is a Win32 version of the libpcap UNIX utility. As of the writing of this document, WinPCap 2.2 does not work with Windows XP; therefore it is necessary to run the beta 2.3 version of WinPCap. After WinPCap has been loaded, WinDump and Ethereal are ready to install.

WinDump is a simple application that is run from a command prompt. Once WinDump has been downloaded, it should be copied to the ⟨%SystemRoot%\ system32⟩ directory so that it can be run from any command prompt. WinDump is good for generating raw packets.

As for Ethereal, it has a GUI that is far more advanced than WinDump. Install Ethereal into a directory of you choice and it is ready to go. Ethereal is good for looking at packets in a decoded mode and is much easier to view packets.

The sniffers that we have discussed so far are only good for sniffing when the client is associated with the AP and for 802.11b frames that are not encrypted with WEP. In a situation where an AP is using a WEP key to cipher its data, it will be necessary to use a different type of sniffer.

AirSnort, a UNIX utility, is a special type of sniffer that will crack the APs WEP key. AirSnort must be run long enough to collect between 500 Megabytes to 1 Gigabyte of traffic in order to retrieve the key. This can take a few hours or significantly longer, based upon network traffic. AirSnort exploiting the undersized 24-bit IV, so it makes no difference if the WEP key is 64-bit or 128-bit.

WEPcrack is a script that can be run against a raw capture file created by Ethereal and it too must also be run on a UNIX system. Ethereal packet captures can be exported to a file and WEPcrack can be used to devise the static WEP key.

The fact that this document is utilizing Windows XP for the penetration test, it is presumed that another laptop running Linux and compiled with either AirSnort or WEPcrack has already cracked the WEP key. Once the WEP key is known, an AP can be treated as any other.

8.13.3 War Driving – The Fun Begins

In order to penetrate a WLAN, an AP must be located. APs are devices that use Radio Frequency (RF) transceivers in the 2.4 GHz range to connect end users in the same RF range. APs bridge wireless end users to the wired network, and are often located *behind* the firewall. Cheap APs or improperly configured APs broadcast frames that contain information about the WLAN and hackers have built utilities to exploit this information. One such hacker utility is called NetStumbler. A laptop armed with NetStumbler will allow intruders to sniff the air for 802.11b frames with the convenience of driving around in their car.

NetStumbler will log information when it passes within the range of an AP, which is approximately 1–350 feet. NetStumbler is supposed to sound an alarm when it sees an AP, but it was not created with XP in mind. However, NetStumbler can be made to annunciate an alarm in Windows XP by taking any desired ⟨. wav⟩ file and renaming it to ⟨ir_begin.wav⟩, then placing the file in the Windows XP ⟨%SystemRoot%\Media⟩ directory. If the root directory does not contain a subdirectory named media, just create one and place the ⟨ir_begin.wav⟩ file there.

Once NetStumbler is executed, it starts sending out broadcast probes at a rate of once per second. If any APs respond to the probe, NetStumbler will alarm and report information extracted out of the 802.11b frames such as SSID, MAC address, channel, signal strength and whether WEP is on. NetStumbler can also be configured to use a GPS to locate the global position of an AP. This is very convenient for pinpointing a certain AP when NetStumbler has discovered many APs in a general area.

NetStumbler is only effective if the AP is responding to broadcast probes and can be made obsolete if the AP is configured to not broadcast the SSID. Many hardware vendors have solutions that can resolve broadcasting issues ranging from shutting off the broadcast to negotiating a broadcast encryption key. It is highly recommended to prevent an AP from broadcasting unless it is encrypted.

8.13.4 The Penetration

Now that an AP has been located, it is time to gather information to see if the AP is vulnerable and welcomes hackers into the LAN. This is where 'Penetration Testing' comes into effect on a WLAN segment.

Some WLAN administrators will set up a DHCP server for the WLAN segment that will assign a wireless NIC an IP address and gateway. If this is the case, an attacker has already successfully gained access to the network. There is nothing more for an attacker to do than begin scanning the network.

If the laptop and wireless NIC are *Associated* with the AP (Layer 2) but do not have an assigned IP address (Layer 3) for the local WLAN segment, they cannot participate on the TCP-IP WLAN. In order to have routing privileges or Internet connectivity, the wireless NIC needs a layer 3 IP address and default gateway. Gaining an IP address can be accomplished with Ethereal or WinDump by sniffing the air medium for packets containing the vital IP information.

The Ethereal GUI can be used to import packets picked up by the Orinoco Gold NIC and decode them for easy viewing. WinDump can be used for the same purpose but it works in a command prompt and visually shows all packets received by the Orinoco Gold NIC as they enter the interface. This will reveal source and destination IP addresses of devices on the WLAN segment.

WinDump can be made to use a specific adapter interface and even dump output to a file. The interface that WinDump is to sniff must be represented by the registry string settings for the desired NIC interface. These wireless NIC registry settings can be conveniently found in Ethereal by hitting 'Ctrl – K' and copying the text in the 'Interface' box for the desired NIC. Here is an example command that allows WinDump to sniff an interface and dump its output to a file called WarDrive.txt.

C:\ >windump -i \Device\Packet_{BAC2F63F-45D5-4AC3-9C3C-73E0ADAE054D}
    ≫WarDrive.txt

After the necessary IP information has been uncover by WinDump or Ethereal, it can be easily applied to the wireless NIC. This fully arms the laptop with a connection to the WLAN and an IP stack to route on the WLAN segment. As can be imagined, this will cause all kinds of problems for an administrator.

Once there is an *Association* with the AP and a proper IP address and subnet mask assigned to the wireless NIC, an attacker can start to probe the network for further layer 3 information. In order to move from the local WLAN segment to other parts of the network, it is necessary to find the nearest gateway router. This can be done with a quick ping scan of the local segment.

Rhino9 Pinger v1.0 is an application that can ping an entire subnet, ping a specific range of IP addresses, and locate all ICMP enabled devices on the WLAN segment. This utility will also resolve the hostnames of the pinged devices. This is very beneficial when it comes to locating the gateway router. If it is not evident which device is the gateway router, just begin to try various IP addresses for the laptops gateway.

A better way to detect the gateway is to scan the newly discovered IP addresses with Nmap and selecting Operating System (OS) detection. Once a router IOS shows up, try the device IP as the laptop gateway. After the gateway router is found and the laptop is configured, verify the IP stack is correct by entering ⟨*ipconfig/all*⟩ in a command prompt.

If the gateway router has a connection to the Internet, then the laptop also has WWW access. This, of course, is only true if there are no firewalls behind the router or a router Access Control List (ACL) to prevent egress to the Internet or other parts of the network. An intruder that has access to the Internet can use the WLAN to download other hacking tools and perform attacks on the local network. The intruder can also attack other networks on the Internet disguising their conduct as the penetrated WLAN.

8.13.5 *Problems caused by Wireless Hackers*

Now that there is full access to the LAN and Internet, an attacker is free to exploit the network for any vulnerabilities or misconfigurations. Nmap is also a terrific port scanner for verifying what ports are open on the discovered IP addresses. This will tell the attacker what type of OS is running, what services are running and what exploits should be conducted next.

For example; let's say the attacker has discovered that the LAN consists of NT Servers. Unless properly configured, the NT machines will allow 'Null Sessions' with their IPC$ shares. By establishing a null session with an NT machine, an intruder can extract extremely critical information from the NT network. Such information can include the Domain name, PDC and BDC info, share names and user accounts. A null session can be achieved by issuing a 'Net Use' command with an empty password in an ordinary command prompt. Here is an example:

C:\>net use\\192.168.0.1\IPC$ '''' /u:''''

Once a Null Session has been executed successfully, an attacker can use hacking tools like NetBIOS Auditing Tool (NAT) to find remote name tables and even crack passwords. NAT allows an intruder to extract various user account information from an NT Server and perform password attacks. Using the extracted usernames to devise username and password dictionary files does this. If an account is set up with a weak password or no password at all, NAT could possibly compromise a user account or even an administrator's account. This is an extremely common situation and has very serious repercussions.

There is a fair chance that NAT will be able to exploit an administrator's password, which will grant the attacker administrative rights for the NT domain. Administrative rights on a domain, in turn, give the hacker the ability to attach to any Microsoft Window machine on the domain or any trusted domain. This includes a range of abilities from deleting Windows NT user accounts to taking a domain controller off line. In short, the attacker is now the networks new and unethical administrator.

L0phtCrack 3.0 (LC3) is a utility that will crack encrypted Windows NT passwords. With the newly acquired administrative rights, a hacker will be able to connect to the PDC with LC3 and withdrawal *all* users accounts and crack *all* passwords on the NT Domain. LC3 is a favorite among hackers and is one of the best password cracking utilities available today. As can easily be seen, once a hacker has compromised the PCD Security Account Manager (SAM), the NT domain is at the will of the intruder.

8.13.6 *Security Recommendations*

As for security on the wireless segment, the WLAN should be regarded as the public Internet and all traffic should be treated as subject.

- Purchase WLAN products that have proprietary security mechanism to overcome the shortcomings of the 802.11b security standards. Many hardware vendors are creating APs that utilize per user and per session WEP keying along with per packet authentication.
- Install RADIUS servers on the wired LAN to aid in the authentication process of WLAN users. Extensible Authentication Protocol (EAP) can be used in conjunction with 802.1X to block traffic to the wired LAN until the RADIUS server has authenticated the WLAN user.
- Place a firewall in front of the AP so all traffic to the wired LAN can be filtered and screened for malicious activities. All services not being utilized should be disabled and logging should dump to a Syslog Host located in a

Demilitarized Zone (DMZ). The Syslog Host will log all incoming traffic and act as a first line of defense in detecting attacks aimed at the router and firewall interfaces. It is also recommended to implement an Intrusion Detection System (IDS).

- Utilize VPN technologies to ensure proper confidentiality, authentication, integrity and non-repudiation of all WLAN usage. This type of environment can incorporate both hardware and software solutions that provide a minimum-security standard of:
  - IKE – 3DES, SHA-HMAC, DH Group 2 and preshared key.
  - IPSec – 3DES, SHA-HMAC, no PFS and tunnel mode.

  Wired network security consists of the same 'good old-fashioned' policies that should be followed every day. It is a best practice to lock down everything, check all IDS logs and keep a constant eye on any up-and-coming exploits.

- All Domain Controllers should make use of the S/Key utility located in Windows NT Service Packs 3 and greater. This utility prevents attackers from remotely retrieving usernames and passwords from domain controllers with LC3.

- Microsoft has created several Security Checklists on how to tighten up and lockdown a Windows NT Domain along with all of its workstations and servers. They consist of several stringent documents and it is highly recommended to complete these Checklists on all the NT nodes in the domain.

- It is a best practice to insure all machines have the latest HotFixes applied. To assist in the mass deployment of HotFixes on all the Windows NT machines in the domain, QChain can be used. QChain is an application that allows multiple HotFixes to be executed on a computer without multiple reboots.

- Microsoft Windows NT accounts with no passwords, passwords that are the same as the username and generally weak passwords should be prevented. This can be done by loading the Windows NT User Manager and changing the password to something more secure such as a combination of letters, numbers and alphanumerical characters. Consider implementing a password filter such as passfilt.dll to enhance password security.

- A switch or router with no password or even a weak password will give an attacker freedom into the network. A weak password can be cracked in a matter of seconds by many different software applications or scripts such as Cisco Auditing Tool. Once an attacker gains access to the password, they can configure the network to route or switch traffic at will. The administrator should change the password to something very difficult such as a random combination of letters, numbers and alphanumerical characters.

- The perimeter routers should be configured with very strict granular ACLs that will disable all unnecessary services and put anti-spoofing measures in place. This is absolutely crucial to the security of a network. All routers should also be configured with a banner warning hackers to stay out of the network. This is important for legal reasons because it will act as an official warning. Too many hackers get off by playing stupid.

8.14 The 802.15 WPAN Standard (Bluetooth)

8.14.1 Overview of the 802.15 WPAN

The Wireless Personal Area Network4 (WPAN) technology uses a short-range radio link that has been optimized for power conscious, battery operated, small size, lightweight personal devices.

The 802.15 WPAN supports both synchronous communication channels for voice-grade communication and asynchronous communications channels for data communications. An 802.15 WPAN is created in an *ad hoc* manner whenever an application in a device desires to exchange data with matching applications in other devices. Once the applications involved have finished their tasks, the 802.15 WPAN will cease to exist.

The 802.15 WPAN operates in the Industrial, Scientific, Medical, or ISM band-width located in the 2.4 GHz range. A fast frequency hopping (1600 hops/sec) transceiver is used to reduce interference and fading in this band. A Gaussian-shaped, binary FSK with a symbol rate of 1 M symbols/s minimizes transceiver complexity. A slotted channel is used, which has a slot-duration of 625 $\propto$s. A fast time-division duplex (TDD) scheme is used that enables full duplex communica-tions at higher layers. On the channel, information is exchanged through packets. Each packet is transmitted on a different frequency in the hopping sequence. A packet nominally covers a single slot, but can be extended up to either three or five slots. For data traffic, a one-direction (asymmetric) maximum of 723.2 kb/s is possible between two devices. A bi-directional 64 kb/s channel supports voice-traffic between two devices. The jitter for the voice traffic is kept low by using small transmission slots.

Figure 8.21 shows the general format of a single-slot, payload-bearing packet transmitted over-the-air in an 802.15-based WPAN. The packet is comprised of a fixed size access code, which is used, among other things, to distinguish one WPAN from another, a fixed size packet header, which is used for managing the transmission of the packet in a WPAN; and a variable size payload, which carries upper layer data. Due to the small size of these packets, large upper-layer packets need to be segmented prior to transmission over the air.

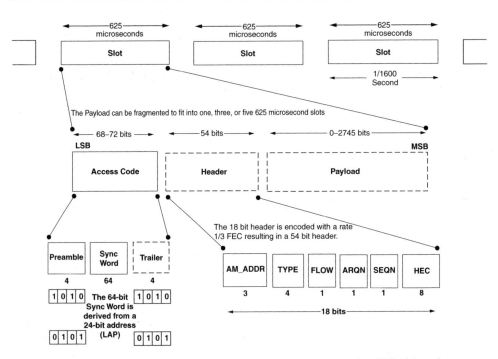

Figure 8.21   The format of an over-the-air payload bearing WPAN packet

8.14.2 High-Level View

The 802.15 standard defines a WPAN that utilizes the Bluetooth wireless technology. In the text of this book, unless otherwise stated, the term 'Bluetooth WPAN' or simply '802.15 WPAN' will refer to a WPAN that utilizes the Bluetooth wireless technology. The term 'Bluetooth wireless technology' and other similar terms will also be used to further emphasize the use of this technology in the Bluetooth WPAN defined and described here.

Wireless personal area networks (WPANs™) are used to convey information over short distances among a private-intimate group of participant devices. Unlike a wireless local area network (WLAN), a connection made through a WPAN involves little or no infrastructure or direct connectivity to the world outside the link. This allows small, power-efficient, inexpensive solutions to be implemented for a wide range of devices.

The essential components of the P802.15 WPAN architecture are based on the Bluetooth Foundation Specification version 1.1. The term 'WPAN' is the name trademarked by IEEE to describe a particular category of wireless communications technology. The Bluetooth wireless technology is an industry specification for small form factor, low-cost, wireless communication, and networking between

PCs, mobile phones, and other portable devices. The proposed 802.15 or 'WPAN' Standard is being developed to ensure coexistence with all 802.11 networks.

Specifically, this standard:

- Describes the functions and services required by an IEEE 802.15 device to operate within *ad hoc* networks.
- Describes the MAC procedures to support the Asynchronous Connection-Less (ACL) and Synchronous Connection-Oriented (SCO) link delivery services.
- The Baseband (BB) layer, specifying the lower-level operations at the bit and packet levels (FEC operations, encryption, CRC calculations, and ARQ protocol).
- The Link Manager (LM) layer, specifying connection establishment and release, authentication, connection and release of SCO and ACL channels, traffic scheduling, link supervision, and power management tasks.
- The Logical Link Control and Adaptation Protocol (L2CAP) layer has been introduced to form an interface between standard data transport protocols and the Bluetooth protocol. It handles multiplexing of higher-layer protocols, and segmentation/reassembly of large packets. The data stream crosses the LM layer, where packet scheduling on the ACL channel takes place. The audio stream is directly mapped on an SCO channel and bypasses the LM layer. The LM layer, though, is involved in the establishment of the SCO link. Between the LM layer and the application, control messages are exchanged in order to configure the Bluetooth transceiver for the considered application.
- Describes the 2.4 GHz ISM band PHY signaling techniques and interface functions that are controlled by the IEEE 802.15 MAC. Requirements are defined for two reasons:
  - provide compatibility between the radios used in the system, and
  - define the quality of the system.

Above the L2CAP layer, RFCOMM, Service Discovery Protocol (SDP), Telephone Control Specification (TCS), Voice-quality channels for audio and telephony, and other network protocols (e.g. 802.2 LLC, TCP/IP, PPP, OBEX, Wireless Application Protocol) may reside.

Personal electronic devices are becoming more intelligent and interactive. Many devices, such as notebook computers, cellular phones, personal digital assistants (PDAs), personal solid-state music players, and digital cameras, have increased data capabilities. This capability allows them to retain, use, process, and communicate various amounts of information. For example, several of these personal devices have a personal information management (PIM) database maintaining personal calendars, address books, and to-do lists. PIM databases in one personal

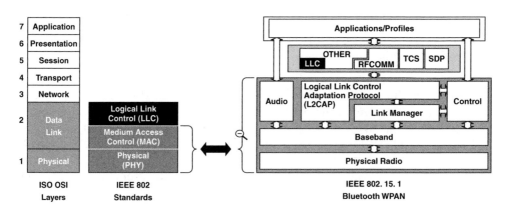

Figure 8.22   Mapping of ISO OSI to scope of 802.15 WPAN standard

device should remain synchronized with PIM databases in other personal devices. The obvious solution for keeping these PIM databases synchronized is to inter-connect and synchronize them.

Traditionally, proprietary special-purpose cables have been used to interconnect personal devices. However, many users find using these cables to be quite a frustrating and unproductive chore. Cables may be lost or damaged and they add unnecessary bulk and weight when carried around. It thus becomes quite desirable to develop connectivity solutions for interconnecting personal devices that do not require the use of cables. Since the desired solution will not involve interconnection cables, a wireless solution must be employed.

Figure 8.22 shows the protocol stacks in the OSI 7-layer model and in the Bluetooth wireless technology and their relation as it pertains to this standard. As shown in Figure 8.22, the logical link control (LLC) and MAC sublayers together encompass the functions intended for the data link layer of the OSI model. The MAC to LLC service definition is specified in ISO/IEC 15802-1:1995.

8.14.3 The General Requirements of 802.15

The demands of the consumer market require connectivity solutions that respect the primary functionality of the personal device. For example, a communications-enabled PDA must still look and function as a PDA. Its wireless connectivity solution must not impact the PDA's form factor, weight, power requirements, cost, ease of use, or other traits in a significant way.

Personal devices used as a part of an individual's productivity management and entertainment tools may also be needed to interact with a corporate IT infrastructure as well. Further, these devices will likely be used in several of many different environments such as offices, homes, in the middle of a park, industrial plants, and so on. The wireless solution for these devices must therefore accommodate the design and marketing requirements dictated not only by the consumer market but by the business market as well.

Interconnecting personal devices is different from computing device connectivity. Typical connectivity solutions for computing devices, like a WLAN connectivity solution for a notebook computer, associates the user of the device with data services available on, for instance, a corporate Ethernet-based intranet. This contrasts with the intimate, personal nature of a wireless connectivity solution for the personal devices associated with a particular user. The user is concerned with electronic devices in his/her possession, or in his/her vicinity, rather than to any particular geographic or network location. The term *personal area network* (PAN) was coined to describe this different kind of network connection. The untethered version of this is called a *wireless personal area network* (WPAN). A WPAN can be viewed as a personal communications bubble around a person. Within this bubble, which moves as a person moves around, personal devices can connect with one another. These devices may be under the control of a single individual or several people's devices may interact with each other.

Communicating devices may not be within line of sight of each other. For this reason WPANs may employ radio frequency (RF) technologies to provide the added flexibility to communicate with hidden devices. This standard presents a WPAN using RF technology based on the Bluetooth wireless technology.

8.14.4 How WPANs differ from WLANs

At first glance, the operation and objectives of a WPAN may appear to resemble those of a wireless LAN (WLAN), like IEEE 802.11. Both the WLAN and WPAN technologies allow a device to connect to its surrounding environment and exchange data with it over an unlicensed, wireless link. However, WLANs have been designed and are optimized for usage of transportable, computing (client) devices, such as notebook computers. WPAN devices are even more mobile.

The two technologies differ in three fundamental ways:

- Power levels and coverage.
- Control of the media.
- Lifespan of the network.

8.14.5 Power Levels and Coverage

To extend the wireless LAN as much as possible, while minimizing the burden of multi-hop networks, a WLAN installation is often optimized for coverage. Typical coverage distances are on the order of 100 meters and are implemented at the expense of power consumption (typically 100 mW or more of transmit power). The added power consumption for covering larger distances has an impact on the devices participating in a WLAN: they tend to be connected to a power plug on the wall, or utilize the wireless link for a relatively short time while unplugged.

A WLAN enables ease of deployment, where the much more desirable (with respect to reliability and bandwidth) cables are very hard or costly to deploy. WLAN links have been designed to serve as a substitute for the physical cables of a LAN cabling infrastructure. While wireless connectivity allows for portability of the client devices, the WLAN itself is quasi-static. A client device in a WLAN will typically be connected to a fixed base station and, on occasion, it may roam between these fixed base stations. While WLANs are much easier to deploy as compared with their wire line counterparts, they still need to be deployed and setup. Their primary orientation is reaching outward from portable devices to connect to an established infrastructure, wired or not.

WPANs are oriented to interconnect multiple mobile, personal devices. The distinction between 'mobile' and 'portable' devices in this discussion is that mobile devices will typically operate on batteries and will have a fleeting interconnection with other devices; portable devices are moved less frequently, have longer time periods of connections, and usually will run from power supplied by wall sockets. A personal device may not necessarily have the need to access LAN level data services, but access to data services on a LAN is not excluded.

In contrast to a WLAN, a WPAN trades coverage for power consumption. Through small coverage area on the order of 10 meters – reduced power consumption – typically 1 mW of transmit power and through low power modes of operation, a WPAN can achieve sufficiently small power-consumption rates to enable portability. Several simple, power-conscious, personal devices can, therefore, utilize a WPAN technology and share data and be truly mobile.

8.14.6 Control of the Medium

Given the high variety of personal devices that may participate in a WPAN, the WPAN technology must support applications with stringent (reserved) bandwidth requirements as well as those with more flexible bandwidth requirements. To provide the necessary bandwidth guarantees for the various connections, the WPAN employs a controlling mechanism that regulates the transmissions of the devices in the WPAN.

WLANs employ similar coordination functions as an option, recognizing that over the large distances covered by them, it may not always be desirable to have a strict and absolute control of the media. When they do have this level of control, it is described as a 'contention free period'. The latter does not mean an interference-free environment. Other independently operating networks (of various technologies) may occasionally interfere with transmissions during a 'contention free period'. However, no contention resolution mechanisms are employed to recover from disturbed transmissions during a contention-free period. With this in mind, in 802.15 WPANs, all the time is 'contention free'. This level of control is achieved by creating a relationship (in 802.15 a master/slave relationship) between the devices and operating on a single, time-multiplexed, slotted system. The 802.15 WPAN master polls its collection of 802.15 WPAN slaves for transmissions, thus regulating the bandwidth assigned to them based on quality of service requirements that it enforces. Using small-sized slots efficiently controls the jitter in transmissions experienced by the high quality traffic. In addition, employing a frequency-hopping scheme with the small sized slots provides noise resilience from interference that may occur by other networks, including other independently operating 802.15 WPANs, operating over unlicensed bands.

The *ad hoc* nature of connectivity in a WPAN implies that devices may need to act as either a master or a slave at different times. As a result, the design objectives for the 802.15 WPAN technology (low-cost, low-power, etc.) still apply no matter whether a device implements the typical master-or-slave capability, or only one of them. In all cases, a master must be present for communications to occur.

Personal devices that participate in a WPAN are designed for their personal appeal and functionality. They are not designed to be members of an established networking infrastructure, even if they may connect to it when necessary. A typical WPAN device does not need to maintain a network-observable and network-controllable state. WLANs are required, for example, to maintain a *management information base* (MIB). As *bona fide* members of a larger infrastructure, this is appropriate. However, with WPAN technology it may be inappropriate, if not impossible, for an end-to-end networking solution to be employed that is observed and controlled remotely over a network. Such end-to-end solutions can be built on top of the WPAN technology and are outside the scope of this standard and will likely be very application-specific.

8.14.7 Lifespan of the Network

In a WPAN, a device creates a connection that lasts only for as long as needed and has a finite lifespan. For example, a file transfer application may cause a connection to be created only long enough to accomplish its goal. When the application terminates, the connection between the two devices may be severed as well. The connections that a mobile client device creates in a WPAN are ad hoc and temporary in nature. The devices to which one's personal device is currently connected in a WPAN may bear no semblance to the device that its was previously connected to or it will connect next. For example, a notebook computer may connect with a PDA at one moment, a digital camera at another moment, or a cellular phone at yet another moment. At times, the notebook computer may be connected with any or all of these other devices. The WPAN technology must be able to support fast, in the order of few seconds, ad hoc connectivity with no need of pre-deployment of any type.

8.14.8 802.15 Security

In order to provide user protection and information confidentiality, the system has to provide security measures both at the application layer and the link layer. These measures need to be appropriate for a peer environment. This means that in each WPAN unit, the authentication and encryption routines are implemented in the same way.

Four different entities are used for maintaining security at the link layer:

- A public address which is unique for each user;
- Two secret keys;
- A random number which is different for each new transaction.

The four entities and their sizes as used in WPAN are summarized in Table 8.1.

The WPAN device address (BD_ADDR) is the 48-bit IEEE address, which is unique for each WPAN unit. The WPAN addresses are publicly known, and can be obtained via MMI interactions or automatically, via an inquiry routine by a WPAN unit.

The secret keys are derived during initialization and are further never disclosed. Under normally circumstance, the encryption key is derived from the authentication key during the authentication process. For the authentication algorithm, the

Table 8.1 Entities used in
authentication and encryption
procedures

Entity	Size
BD_ADDR	48 bits
Private user key, authentication	128 bits
Private user key, encryption configurable length (byte-wise)	8–128 bits
RAND	128 bits

size of the key used is always 128 bits. For the encryption algorithm the key size may vary between 1 and 16 octets (8–128 bits). The size of the encryption key shall be configurable for two reasons.

(1) The result of the many different requirements imposed on cryptographic algorithms in various countries.
  – Both w.r.t. export regulations and official attitudes towards privacy in general.
(2) To facilitate a future upgrade path for the security without the need of a costly redesign of the algorithms and encryption hardware. Increasing the effective key size is the easiest way to address increased computing power at the opposite side.

Even though the authentication key is used when creating the encryption key, the two are totally different from each other. Each time encryption is activated, a new encryption key is generated. Therefore, the lifetime of the encryption key does not necessarily correspond to the lifetime of the authentication key.

It is anticipated that the authentication key will be more static in its nature than the encryption key. Once the authentication key is established then it is up to the application running on the WPAN device to decide when to change. To underline the fundamental importance of the authentication key to a specific WPAN link, it will often be referred to as the link key.

The RAND is a random number that is derived from a random or pseudo-random process in the WPAN unit. This is not a static parameter since it changes frequently.

8.14.9 *Authentication*

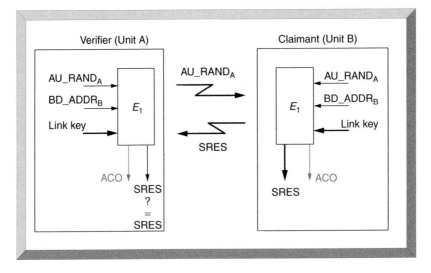

Figure 8.23   Challenge-response for the WPAN

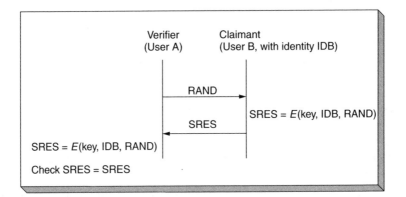

Figure 8.24   (Siep, 2000) Challenge-response for symmetric key systems

The authentication routine used in WPANs uses a challenge–response scheme (Figure 8.23) in which a claimant's knowledge of a secret key is checked through a two-move protocol using symmetric secret keys (Figure 8.24). The latter implies that a correct claimant/verifier pair shares the same secret key, for example K. In the challenge–response scheme the verifier challenges the claimant to authenticate a random input (the challenge), denoted by AU_RAND_A, with an authentication code, denoted by E_1, and return the result SRES to the verifier, see Figure 4.2. This figure shows also that in WPAN the input to E_1 consists of the tuple AU_RAND$_A$ and the WPAN device address (BD_ADDR) of the claimant. The use of this address prevents a simple reflection attack. The secret K shared by units A and B is the current link key.

In the WPAN, the verifier is not necessarily the master. The application indicates who has to be authenticated by whom. Certain applications only require a one-way authentication. However, in some peer-to-peer communications, one might prefer a mutual authentication in which each unit is subsequently the challenger (verifier) in two authentication procedures. The LM coordinates the indicated authentication preferences by the application to determine in which direction(s) the authentication(s) has to take place. For mutual authentication with the units of Figure 8.23, after unit A has successfully authenticated unit B, unit B could authenticate unit A by sending a AU_RAND$_B$ (different from the AU_RAND$_A$ that unit A issued) to unit A, and deriving the SRES and SRES' from the new AU_RAND$_B$, the address of unit A, and the link key K_{AB}.

If an authentication is successful, the value of ACO as produced by E_1 should be retained.

Note that during each authentication, a new AU_RAND$_A$ is issued. Mutual authentication is achieved by first performing the authentication procedure in one direction and immediately followed by performing the authentication procedure in the opposite direction. As a side effect of a successful authentication procedure an auxiliary parameter, the Authenticated Ciphering Offset (ACO), will be computed. The ACO is used for ciphering key generation. The LM determines the claimant/verifier status.

When authentication attempts fail, a waiting interval passes before the verifier will initiate a new authentication attempt to the same claimant or before it will respond to an authentication attempt initiated by a unit claiming the same identity as the suspicious unit. For each subsequent authentication, failure with the same WPAN address the waiting interval shall be increased exponentially. This means that after each failure the waiting interval before a new attempt can be made is twice as long as the waiting interval prior to the previous attempt. To make everyone rest at ease, the waiting interval is limited to a maximum. The maximum waiting interval is dependant upon the initial implementation. The waiting time is exponentially decreased to a minimum when no new failed attempts are being made during a certain time period. This procedure prevents an intruder from repeating the authentication procedure with a large number of different keys. In order to make the system a little less vulnerable to denial-of-service (DoS) attacks, the WPAN units need to keep a list of individual waiting intervals for each unit it has established contact with. Because it is possible to end up with a very large list, the size of this list must be restricted only to contain the N units with which the most recent contact has been made. The number N can vary for different units depending on available memory size and user environment.

8.15 The 802.16 Standard

IEEE Standard 802.16 – the Institute of Electrical and Electronic Engineers Standards Association's standard board approved Air Interface for Fixed Broadband Wireless Access Systems – on December 7, 2001.

The approval will 'set the stage' for more widespread deployment of 10 GHz to 66 GHz wireless metropolitan area networks, the group says. The interface standard is the first broadband wireless access standard from an accredited standards organization, it says. The standard will be published by January.

The 802.16 standard was created in a two-year, open consensus process involving 'hundreds of engineers'. The standard allows interoperability among products from multiple manufacturers and includes a medium access control layer supporting multiple physical layer specifications, the board says. Extensions to the 2 GHz to 11 GHz bands should be done next summer.

8.16 Mobile IP

8.16.1 The Security of Mobile IP

The purpose of this section is to discuss the security aspects of Mobile IP. Wireless devices such as PDAs (Personal Digital Assistants), Palmtop Computers, even notebook and Laptop computers are now offering IP (Internet Protocol) connectivity. When Mobile IP is used, the connection establish by the wireless devices remains intact.

Just as mobile or cellular phones evolved from the original wireline telephone system It does this without human intervention and non-interactively. The vocabulary normally used to describe this mobile computing is Mobile IP. The original set of Request for Comments (RFC) describing this model is RFCs 2002 through 2006.

The Internet Engineering Task Force (IETF) is a large open international community of network designers, operators, vendors, and researchers concerned with the development of the Internet architecture and the smooth operation of the Internet. RFC 2002, proposed by a working group within the IETF, allows the mobile node to use two IP addresses: a permanent home address and a care-of address that changes at each new point of attachment.

Probably the most important requirement of Mobile IP is to permit a mobile node to communicate using only its home address while varying its point of connectivity to the Internet. Additionally, a mobile node must impart solid authentication when it updates its home agent of its current location.

Mobile IP defines three functional areas:

- A Mobile Node
- A Home Agent
- A Foreign Agent.

A *mobile node* is a host or router that can vary its location from one link to another without changing its IP address and without disrupting ongoing communications.

A *home agent* is a router with an interface on a mobile node's home link that captures packets destined to the mobile node's home address and tunnels them to the mobile node's latest care-of address. The home address is an IP address that is given for an extended period of time to a mobile node. It remains unchanged regardless of where the node is connected to the Internet. The care-of address is the termination point of a tunnel toward a mobile node for datagrams passed on to the mobile node while it is away from home. The protocol can use two different types of care-of addresses. A 'foreign agent care-of address' is an address of a foreign agent with which the mobile node is registered, and a 'co-located care-of address' is a local address that the mobile node has linked with one of its own network interfaces.

A *foreign agent* is a router with an interface on a mobile node's foreign link that helps a mobile node with movement detection and provides routing services on behalf of mobile nodes, including de-tunneling of encapsulated packets when the mobile node uses the foreign agent's care-of address.

Home and foreign agents are typically software products that run on traditional routers or in host computers. Mobile nodes are most common in host computers that are portable, such as notebooks or laptops. It is usual for a single node to be a foreign agent for some mobile nodes while concurrently being a home agent for other mobile nodes.

Home and foreign agents advertise their existence on any attached links by periodically multicasting or broadcasting special Mobile IP messages called Agent Advertisements.

Mobile nodes receive these advertisements and inspect their contents to determine if they are connected to their home link or a foreign link. A mobile node connected to a foreign link gets a care-of address.

A foreign agent care-of address can be read from one of the fields within the foreign agent's Agent Advertisement. The mobile node registers the care-of address with its home agent, using message-exchange defined by Mobile IP. In the registration development, the mobile node requests service from a foreign agent – one is present on the link.

The home agent or some other router on the home link advertises reachability to the network-prefix of the mobile node's home address, thus drawing packets that are intended for the mobile node's home address. The home agent captures these packets and tunnels them to the care-of address that the mobile node registered earlier. At the care-of address, the original packet is removed from the tunnel and then sent to the mobile node. In the reverse direction, packets sent by the mobile node are routed directly to their target, without any requirement for tunneling. The foreign agent serves as a router for all packets created by a visiting mobile node.

9

RFID

9.1 Introduction

Radio Frequency Identification (RFID) Technology is about to enter a boom phase. Whereas in the past its progress was limited due to the lack of technological cost-effective solutions, and therefore served largely specialized niche markets, recent developments now allow for the remaining hurdles to be overcome.

9.1.1 What are RFID Systems?

RFID stands for *radio frequency identification*. It is a widely varied collection of technologies for various applications, ranging from the high-speed reading of railway containers to applications in retail that can be regarded as a potential successor to the bar-coding technologies in use today. RFID is based around radio or electromagnetic propagation. This has the ability to allow energy to penetrate certain goods and read a tag that is not visible thereby to identify those goods remotely, either in the form of an identity code or more simply that something is present (EAS). Different frequencies of the radio system result in different reading ranges and properties of the system.

RFID systems generally comprise two components, namely transponders that are attached to the goods to be labeled, and readers for reading the identity of the transponders. In some cases the transponders might be programmed to broadcast data representing their identity, while in others it might simply be an ON/OFF state such as is used in Electronic Article Surveillance (EAS) systems commonly used for anti-shoplifting in retailing.

Commonly available tags have an operating frequency in the range from 60 kHz to 5.8 GHz depending on application.

Wireless Data Technologies. Vern A. Dubendorf
© 2003 John Wiley & Sons, Ltd   ISBN: 0-470-84949-5

In operation one can generally say that there are three different types of technologies being implemented. They are:

- Magnetic-based RFID technologies
- EAS-based technologies
- Electric field-based RFID technologies.

9.1.2 EAS Systems

EAS stands for Electronic Article Surveillance. Becoming more common in the retail industry nowadays, the EAS systems are used to electronically detect goods that have not been authorized when they are removed from a retailer. The systems comprise a tag attached to the goods and a sensor mechanism. The retailer can neutralize the tag when he wishes to authorize the removal of the goods, for example when the items have been legitimately purchased. In effect, EAS systems are single bit RFID systems, able to convey their presence, but not having sufficient data capabilities to convey an identity.

Presently there are four major technologies used for EAS systems. They are:

- Microwave
- Magnetic
- Acousto-Magnetic
- Radio Frequency.

Market penetration is currently estimated at 6000 million tags per annum at $0.12 each.

The different EAS technologies have widely differing performance in the issues of price, reading range and reliability. The Magnetic and radio frequency versions are very cheap and are generally attached permanently to the goods or their packaging, while the microwave tags are expensive and are removed by the store personnel when the item is paid for, using a special removal tool.

Markers that are left on the goods and neutralized by the sales staff are called deactivatable.

One type of deactivatable marker is in the form of an electronic circuit comprising inductance and capacitance elements that resonate at radio frequencies.

Another type of marker – a magnetic marker – comprises a strip of soft magnetic material that interacts with a ferromagnetic element made of a hard magnetic material that can be magnetized or demagnetized. The soft magnetic strip resonates and generates harmonics in the presence of a magnetic field having a certain frequency. This allows the marker to be identified. The hard ferromagnetic element can be magnetized or demagnetized thereby deactivating or activating the marker.

Another type of marker is the acousto-magnetic or magneto-mechanical marker. This type of marker comprises a strip of magnetostrictive material and a strip of magnetic material of high coercivity. The magnetostrictive material resonates mechanically in the presence of a magnetic field of a particular frequency. A receiver sensitive to the magnetic field created by the mechanical resonating magnetostrictive material can detect this resonance. Modifying the magnetic bias of the strip of magnetic material ordinarily deactivates the marker.

The above systems are commercially available from many competing suppliers.

EAS is a simple addition to electronic RFID systems whose developments have been announced but are as yet still not commercially available. The advantage of such systems with regard to EAS, is that:

- They would broadcast not just the presence of the item triggering the alarm system, but the actual identity of the product.
- They would be turned on and turned off by command allowing the same tagging system to have application at all stations from the manufacturer, through the distribution channels, to the retailer.
- Controlling the tag would not be conspicuous, being incorporated into the reading protocol, rather than the terrible magnetic pads currently used by some retailers that wipes the information from wayward credit cards.
- As the system uses radio communications, the tags can be packaged inside the goods preventing the goods from being removed while the boxes with the conventional EAS tags remain behind in the store.
- The EAS features are incorporated in the identification and tracking system for virtually no additional cost.

These systems are still in their infancy and have a long development path ahead.

9.1.3 Multibit EAS Tags

The following article explains some experimental concepts in achieving multibit or multistatus from a modification of standard EAS techniques. Generally EAS tags are single bit devices and are not switchable in both the on and off direction using a programming signal.

A resonant circuit is one in which the values of circuit resistance, R, capacitance, C, and inductance, L, are chosen such that the reactance of the resonant circuit is a minimum at a resonant frequency.

One method that is used is for a resonant circuit to be disposed on a thin insulating dielectric substrate to form a tag for use in electronic article detection (EAS) schemes. Generally, the coil of the resonant circuit consists of a closed loop of a conducting element that has a certain value of resistance and inductance. A capacitive element that forms part of this closed loop consists of two separate areas of thin metal conducting film disposed on opposite sides of the dielectric. The tag is attached to articles to be protected from theft. An RF signal at or near the resonant frequency of the resonant circuit is emitted from a base station. When the tag is in the RF field, the tag's absorption can lead to a change in the tank circuit current of the base station and a power dip in a receiving coil. Both of these effects can be used to sense the presence of the tag and hence the item to which it is attached. Thus, an alarm can be made to sound when either of these effects are sensed by a pickup coil or by an amplifier, indicating improper removal of an item. To deactivate the tag, a relatively high RF power pulse can be applied at the counter at which the point-of-sale of the item takes place. This high power acts to short the capacitor or burn out a weak portion of the coil. In either case, the circuit is no longer resonant and will not respond to the RF interrogation from the base station. Therefore, the customer who has made a legitimate purchase at the point-of-sale counter can pass through the interrogation-sensing gate without setting off an alarm.

It is clear from this description that these tags, once deactivated, are not reusable. In addition, in the configuration just described, the tags are capable of only conveying one bit of information. Thus, they cannot give any information regarding the item's identification and are useful only for anti-theft applications. This kind of tag is normally classified as a single bit tag.

Some RF tags consist of a resonant coil or a double-sided coil containing two thin film capacitors with the plate of each capacitor on opposite sides of the dielectric. Such tags can be used for source tagging and have an initial frequency that is different from the frequency used at the retail establishment for theft protection. For example, the tag is designated as being in a deactivated state until the first capacitor is shorted by means of a high power RF pulse at the resonant frequency. Disabling the capacitor shifts the resonant frequency of the RF circuit to the store interrogation frequency. A second deactivation pulse is used to disable the second capacitor at the point-of-sale when payment is received for the item to which the tag is attached. At this stage, the tag is no longer usable and has been permanently destroyed.

Some other systems have been proposed where two or more frequencies can be obtained on an RF coil tag by altering the capacitance of the circuit. In one case, a strong DC electric field is applied to change the effective dielectric constant of the capacitor. Thus, the circuit has two resonant frequencies depending on the value of the applied electric field. Due to the ferroelectric hysteresis, the tag can be deactivated by the application of a DC field. However, it can also be reactivated and hence re-used by applying a DC field of opposite polarity.

In another version, a set of capacitors connected in parallel attached to an inductance have been described in which each dielectric of the set of capacitors varies in thickness. In this manner, a series of resonant frequencies can be obtained by applying different voltages (electric fields). Each of the capacitors then changes capacitance at a different electric field (voltage) levels depending on the thickness of the dielectric.

Another concept consists of an array of series capacitors connected in parallel with an inductor. Here, selectively shorting one or more of the capacitors, thereby changing the resonant frequency of the resulting circuit, can alter the resonance. A frequency code can thereby be established by disabling or burning out selective capacitors at the time of interrogation, those capacitors becoming disabled which at the time of manufacture of the tag were 'edimpled'. The tag is not reusable once scanned since the code relies on burning out a capacitor during the scan cycle and observing the change in frequency. Thus, once the tag has been queried its capacitive elements become irreversibly shorted and hence the tag cannot be scanned again.

An idea for a reusable tag comprises of two ferromagnetic elements, one soft (low coercivity) and one hard (high coercivity) both physically covering a portion of an RF coil. The ferromagnetic element with high coercivity can be magnetized to apply a bias field to the soft material to put the latter into saturation. In that state, the RF field generates very small hysteresis losses leading to a relatively high Q of the tag circuit. On the other hand, when the hard magnet is demagnetized, the RF field results in hysteresis losses in the soft material that lowers the Q of the circuit. This change in Q can be used to determine whether a tag is active or has been deactivated.

A reader apparatus for interrogating and sensing the presence of an RF resonant tag is realized where the interrogating frequency is swept around a center frequency. In general, there is very little radiation emitted except when the tag is present in the field of the emitter. Thus, when there is no tag in the antenna field, very little energy is lost from the antenna circuit. When the swept frequency coincides with the resonant frequency of an active tag, energy is absorbed and a sensing circuit detects a drop in voltage level in the interrogating antenna oscillator circuit. The tag absorption occurs twice with every complete sweep cycle resulting in a negative dip in the oscillator circuit. The negative dip causes pulse modulation that is filtered, demodulated and amplified to cause an alarm to be activated, indicating theft of an item. Thus, the basic detection is achieved by varying the interrogation carrier frequency to match the resonance of a tag whose center frequencies span a range depending on the type or make of tag.

Retail tagging, tagging used in the road/air-freight package industry, personnel identification tagging, pallet tagging in manufacturing processes, etc., requires a tag for identifying a product, article or person in detail. With a sufficient number of bits, the tag can be interrogated to yield useful information such as what the product is, its date of manufacture, its price, whether the product, article or person has been properly passed through a check-out counter or kiosk, etc. Further, identifying a large number of products via tags can lead to a new type of check-out system for the retail industry giving rise to the much hoped for 'no-wait check-out'.

Conventional tags and tag systems have had a number of problems including: (1) having only one bit, typical of anti-theft tags, or (2) requiring a large amount of power to read the tag, thus requiring a tag battery (or other suitable power source), or (3) being relatively easy to defeat by tampering.

Multibit, remotely sensed tags are needed for retailing, inventory control and many other purposes. For many applications, the cost must be low and the tags must be able to be individually encoded. Further, when the tag is interrogated it must produce a distinctive signal to reliably identify the article to which the tag is attached or coupled.

Some conventional tags have employed the Barkhausen jump effect. Generally, the Barkhausen effect is characterized by a tendency for magnetization to occur in discrete steps rather than by continuous change, thereby giving rise to a large temporal flux change, $d\phi/dt$, which is key for inducing a sizable voltage in a sensing or pickup cot.

For example, US Patent No. 5 181 020 describes a thin-film magnetic tag having a magnetic thin film formed on a polymer substrate and a method for producing the same. The thin film exhibits a large Barkhausen discontinuity without intentional application of external torsion or tensile stress on use. A particular disclosed use is as a marker or tag for use in an article surveillance system wherein articles may be identified by interrogating the tagged article in a cyclic magnetic field of a predetermined frequency in a surveillance area and detecting a harmonic wave of the magnetic field generated by the tag in the surveillance area.

This conventional system is only a single bit element using a single Barkhausen layer with no ability to develop a code to distinguish items.

US Patent No. 5 313 192 describes another single bit tag that relies on the Barkhausen effect. The tag of this invention is selected to include a first component comprised of a soft magnetic material that constitutes the bulk of the tag. A second component comprised of a semi-hard or hard magnetic material is integral with the first component. The tag is conditioned such that the second component has activating and deactivating states for placing the tag in active and deactivated states, respectively. Such conditioning includes subjecting the composite tag to predetermined magnetic fields during thermal processing stages.

By switching the second component between its activating and deactivating states the tag can be switched between its active and deactive states. A reusable tag with desired step changes in flux that is capable of deactivation and reactivation is thereby realized.

US Patent No. 4 980 670 describes a one-bit magnetic tag formed from a magnetic material having domains with a pinned wall configuration. The resulting hysteresis characteristic for that material is such that upon subjecting the material to an applied alternating magnetic field, the magnetic flux of the material undergoes a regenerative step change in flux (Barkhausen jump) at a threshold value when the field increases to the threshold value from substantially zero and undergoes a gradual change in flux when the field decreases from the threshold value to substantially zero. For increasing values of applied field below the threshold, there is substantially no change in the magnetic flux of the material. The tag may be deactivated by preventing the domain walls from returning to their pinned condition by, for example, application of a field of sufficiently high frequency and/or amplitude.

US Patent No. 4 940 966 describes the use of a plurality of magnetic elements in predetermined associations (e.g. with predetermined numbers of magnetic elements and with predetermined spacings between said elements), for identifying or locating preselected categories of articles. When the articles are caused to move relative to a predetermined interrogating magnetic field, each particular association of magnetic elements gives rise to a magnetic signature whereby the article or category of article carrying each of the predetermined associations can be recognized and/or located.

US Patent No. 4 660 025 describes a marker for use in an electronic surveillance system. The marker, which can be in the form of a wire or strip of magnetic amorphous metal, is characterized by having retained stress and a magnetic hysteresis loop with a large Barkhausen discontinuity. When the marker is exposed to an external magnetic field whose field strength, in the direction opposing the instantaneous magnetic polarization of the marker, exceeds a predetermined threshold value, a regenerative reversal of the magnetic polarization of the marker occurs and results in the generation of a harmonically rich pulse that is readily detected and easily distinguished.

US Patent No. 5 175 419 describes a method for interrogating an identification tag comprised of a plurality of magnetic, thin wires or thin bands that have highly rectangular hysteresis curves and different coercive forces. The wires or bands are preferably of amorphous material, but means for obtaining the highly rectangular hysteresis curves and different coercive forces are not taught; nor is the concept taught of using a time varying magnetic field superimposed on a ramp field for interrogation.

Their invention is an inexpensive multibit magnetic tag which uses an array of amorphous wires in conjunction with a magnetic bias field. The tag is interrogated by the use of a ramped field or an ac field or a combination of the two. The magnetic bias is supplied either by coating each wire with a hard magnetic material which is magnetized or by using magnetized hard magnetic wires or foil strips in proximity to the amorphous wires. Each wire switches at a different value of the external interrogation field due to the differences in the magnetic bias field acting on each wire.

9.1.4 Summary of Limitations of RFID Technologies in their Current State of Development

Except for some recent developments that have still to arrive on the marketplace, transponders technologies have some major millstones around their necks:

- The magnetic-based solutions have limited range, typically a few centimeters and in some cases ranges of about 1 meter.
- The magnetic-based solutions need to operate over short ranges, as they are not generally suitable for the situation where many transponders are in the read zone at the same time.
- The electric field RF transponders have range, but with only a single reading channel that needs to be allocated by a regulatory authority, which often also has the problem of many transponders replying at the same time and causing confusion. Their range, however, can be many meters. The regulatory authorities in the different countries are not able to allocate the same frequency worldwide due to other commitments, and this rules out the facility of onboard receivers should there be a need for world trade.
- The sophisticated warehousing tagging systems have good range and can even be triangulated to provide location, but are unlikely to be a bulk solution due to their high price.
- The EAS technologies are limited in range, and have problems with reliability often due to environmental interference. The two major technologies are both well used, but once deactivated it is not easy to take them back into stock.

The RFID-type technologies are available in many different varieties. Examples of choices are, amongst others, method of energy coupling, operating range, delivered price, singe/multiple targets in a zone, EAS features and price.

9.1.5 *What are Transponders?*

Transponders were originally electronic circuits that were attached to some item whose position or presence was to be determined. The Transponder functioned by replying to an interrogation request received from an interrogator, either by returning some data from the transponder such as an identity code or the value of a measurement, or returning the original properties of the signal received from the interrogator with virtually zero time delay, thereby allowing ranging measurements based on time of flight. As the interrogation signal is generally very powerful, and the returned signal is relatively weak, the returned signal would be swamped in the presence of the interrogation signal.

The functioning of the Transponder was therefore to move some property of the returned signal from that of the interrogation signal so that both could be detected simultaneously without the one swamping the other. The most common property to change is the transmission frequency meaning that the transponder might receive the interrogation frequency at one frequency, and respond on another frequency that is separated sufficiently with regard to frequency so that both may be detected simultaneously.

Transponders were initially used in World War II on aircraft to identify the aircraft using IFF (Identify Friend or Foe), where friendly aircraft would respond to secret preprogrammed interrogation codes and indicate to the radar operators that they were friendly aircraft. Today Transponders are still used extensively on commercial aircraft to relay to the radar operators the height and identity of the aircraft on their radar displays.

Another important use for transponders has been in the measurement of distance. Here the interrogator sends a signal to the transponder, which immediately responds on another frequency. By measuring the time from the sending of the initial signal by the interrogator, to the receipt of the signal from the transponder, and calculating the effective double path traveled using the speed of light, the distance between the transponder and the interrogator can be determined. The accuracy of such systems is limited to fractions of a meter using electromagnetic propagation systems due to the limits in determining the transmission times with sufficient accuracy. (A system called Tellurometer invented in the 1960s improved this resolution over distances of hundreds of kilometers to a few centimeters, but although this still used transponders, it was not based on the principle of time of flight).

Another major category of Transponders that is not the subject of this book is the use of transponders in radio relay systems such as fixed/mobile radio networks and satellite transmissions. The same principle applies in that the data is transmitted on a carrier frequency at one frequency, and rebroadcast on a carrier of another frequency, allowing the strong and weak signals to co-exist.

Transponder systems have recently started to become major players in the field of electronic identification. Within this application, it is necessary to make the transponders as cheap as possible, and to rather build the sophistication into the readers. This lack of sophistication generally means that changing the transmission frequency is no longer an option, as the frequency translation needs expensive and complex tuned circuitry. Instead the transponders have given up the ranging ability and rather time slice the communications channel with the interrogator. Here the interrogator (called a reader) sends an interrogation signal for a limited time. The transponder receives the signal and waits for its completion, and then responds on the same frequency with its identity and data code. (There are more complex methods but these covers the basics.)

The devices are sometimes called transponders and are also sometimes called *tags*, most probably because their end application eventually will be the tagging of goods.

Transponders vary in selling prices from US$1000 down to US$0–20, depending on application and features.

To date there have been some major market developments utilizing transponder technology.

9.1.5.1 Aircraft Identification

Most commercial and general aviation aircraft operating today are fitted with transponder systems. These transponders respond to the air traffic controls secondary radar, providing an identity code that is linked to the echo, enabling the code to be converted to aircraft details on the air traffic controllers (ATC) screen. In addition to an identity code, the Class C type transponders also broadcast the current altitude of the aircraft determined generally from an internal air pressure sensor in the aircraft. Codes to be used for each flight are issued by the ATC to the pilot before the flight, and can be between 0000 and 7777. Certain codes are also reserved for different emergency conditions, enabling the ATC's radar to recognize an aircraft in emergency and bring it immediately to the operator's attention.

9.1.5.2 Railcar and Shipping Container Identification

After aircraft, this phase has most probably been one of the most successful phases of RFID systems. The railway companies have a major system benefit in that they know exactly where coaches using their system will pass the reading point as the coaches are bound to the railway lines. The rail companies have long wanted a system to read the identity of railcars, both for monitoring the integrity of the train as it passes the reading point (avoid the disaster of a full speed train hitting an unexpected stationary coach on what was supposed to be a clear line); and for sorting the coaches when they get reconstituted into different trains at marshalling yards. Initial efforts tried included an ultrasonic and radar-ranging system that read the location of different bars mounted on the side of the coach, but these were replaced with an RFID system. The US company Amtech seems to be the leader in this technology and now has systems fitted to most railcars and rail containers around the world. The transponders are typically 30 cm by 7 cm and are about 1 cm deep. They are usually mounted on the topside of the container, and allow a reader mounted alongside the rail line to read the contents of a train at full speed. (The largest supplier claims to have fitted four million transponders to rail containers so far).

9.1.5.3 Animal Identification

Activity in this area seems to have started in the area of racehorse identification. The intention was to develop a very small transponder that could be hermetically sealed in a tiny glass capsule, and imbedded under the skin of the animal providing positive proof as to the identity of the animal. The transponders are about 10 mm long, 1.5 mm in diameter, and contain a coil inside the glass vial with 1000 turns of very fine wire around a ferrite former. It is this coil that makes the transponders quite difficult to mass produce, and recent adverts indicate prices of $5 each in quantity, with reading ranges of a couple of inches. From race-horses the applications developed first into labeling farm animals and later into labeling household pets. The problem with labeling of the farm animals arose in that the transponder was found to move under the skin and the health authorities were worried that it might get lost and eaten. This created a market for ear tag labeling for farm animals. Many players have now also moved into this market, the most well known being the TIRIS tag from Texas Instruments. Other major manufactures catering to this market are Destron, Trovan and AVID.

Prior to 1997, ear tags that combined a visual reference with an RFID transponder had usually labeled cows. A new series of RFID tags have recently been developed called Ruminary tags, tags that are encased in a tough plastic case which can be fed to the cow with its food, and will reside in the stomach of the cow for the life of the animal. These tags are resistant to attack by the acids in the stomach and operate using magnetic coupling techniques with readers outside the animal.

9.1.5.4 Sports Timing

The most successful transponder systems in sports timing seem to be those used in Grand Prix cars. The system particularly needs to cater for the situation when many cars might be crossing the start line at the same time. The reader is linked to a series of antenna under the track, and the transponders are mounted in the nose of the cars just above the track. Just before the car reaches the start line, it encounters an energy field that 'wakes up' the transponder. The next signal the transponder receives is a synchronizing pulse that is sent out by the reader periodically. The transponders are all allocated a unique time slot to reply in after this synchronizing signal and radiate energy in that slot. By monitoring the timing of the received energy relative to the synchronizing signal, the reader is able to determine all cars present at the time of that synchronizing signal. However, as the RF energy fields (typically 400 MHz) are ill defined, it is necessary for the system to be collocated with a light beam across the track to provide timing accuracy to 1/1000 of a second. This system can only cater for a limited number of transponders (say 64), as each needs to be allocated a unique time slot.

Generally, the timing accuracy provided by transponder systems for sporting events is below the accuracy required by the organizers. This is because the definition of the RF field depends on many different parameters, many not under the control of the designers. This means that the transponder can be used for identification, but time measurement will need to come from other systems.

Most probably the major goal in sports timing in the future will be the timing of road running events. Some of these events have upwards of 40 000 competitors with few organizers to run the events, and yet timing accuracies of only one second are needed, but many competitors can be crossing the line at the same time.

Recent developments in the timing of road races have highlighted a novel application of RFID technologies. In the Boston Marathon (USA) (40 000 runners), Comrades Marathon (South Africa) (13 500 runners) and the marathon race in the Atlanta Olympics, a supplier of transponder systems has been testing a timing system for logging the passage of some of the participants. The system consists of a small magnetic coupled transponder (yellow disc) that is attached to the runner's shoe, and magnetic sensors mounted under carpets that are placed at the start and finish and along the route at strategic locations. The system is able to detect many runners crossing the mats at the same time. Some recent developments in transponder technology using electric field coupling will mean that other cost-effective solutions to this problem are in the pipeline and should see the light of day in the next few years.

9.1.5.5 Toll Road Control

The goal is to electronically identify vehicles passing a toll station at high speed and to debit their accounts automatically for using the toll road.

A variety of different toll road systems are under development around the world, most probably with 80 different manufactures at present. In the next couple of years many different authorities are introducing such systems on a trial basis. The major difficulty in having a national system is the lack of a national standard, as each manufacturer develops their unique solution. The railroad tracking company, Amtech, have for a long time been operating their own toll road to develop their technology. In most of the previous applications, RFID technologies are used in closed solutions, which is the same manufacturer supplies the transponder and the reader. For toll roads, although many users use the same stretch of road every-day, many only occasionally use that road and therefore they are not likely to have transponders fitted. Many users might pass toll locations of different companies that operate with different transponder protocols having numerous different transponders fitted. Basically, without a single standard, the toll road system is likely to be a mess.

A toll road system generally comprises a transponder and a reader. As the vehicles will be passing at high speed, and as the size and shape of the vehicles differ between manufacturers, a reading range of say 4 meters minimum would be required. Ideally, reading speed should also be high. This implies that the toll road system will need an electric field coupled system, say operating at between 900 MHz and 5800 MHz. Transponders are usually mounted in the front window of the vehicle. As of April 1997 it is estimated that there are more than 1.8 million toll road transponders already in use.

9.1.5.6 Electronic Vehicle Key

This is one of the most recent developments in RFID that is starting to take off. Largely due to the inability of the small injectable magnetic transponders to find a market niche in the marking of slaughter animals, the manufacturers have been seeking new markets. As they had already developed a transponder that was small and could fit easily into the plastic base molding of a motor key, they just needed to develop readers that could sense the electronic identity of the key when it was inserted in the ignition. Using the electronic verification, the cars central computer could easily be disabled/enabled to prevent hot wiring. New laws in Europe will soon force cars to have immobilizer systems fitted as standard components, although it is not yet likely that they will be based on RFID technology.

9.1.5.7 Electronic Article Surveillance Systems

EAS systems are commonly seen these days in many retail outlets. Their primary purpose is to discourage shoplifting. The system is a basic RFID system where the information content is generally a single bit, either ON or OFF. The systems comprise a marker label that is attached to the goods to be protected, and a scanning system at the entrance of the store. The labels are all activated on installation, and will be detected by the scanning system if the label in its active state is brought near the scanner, at which stage an alarm is sounded. The labels are either removed (typical for a clothing store) by the sales staff using a special removal tool, when the goods have been bought, or applying magnetic fields or physically destroying the electrical properties of the label neutralizes the labels. EAS systems are widely used and it is estimated that 6 billion units are consumed annually. The EAS is based on properties of magnetic materials that have been known since the 1930s.

9.1.5.8 Four Different Standards are in Widespread Use

Acoustomagnetic

Based on the principles that by exciting a strip of amorphous magnetic material that is also mechanically stressed, with a low frequency magnetic field, the strip will emit harmonics of the scanning signal allowing the harmonics to be detected and hence provide a detection system. Changing the magnetic structure neutralizes the tag.

Electromagnetic

The tag consists of a magnetic material that is illuminated by a pulse of energy, and the decay of the energizing field is monitored. Any responsive magnetic material will modify the decay rate allowing detection of the tags.

Medium RF

The tags comprise a tuned circuit that is tuned to the frequency of the scanning system, typically between 6 MHz and 10 MHz. Printing conductive circuitry on either side of a plastic film, and interconnecting the two surfaces generally make the tags. The label resonates at the scanning frequency when in the presence of the reader. On purchase, the deactivator provides a pulse of energy that overloads the rating of the label and physically destroys the circuit, making it no longer resonant at the scanning frequency.

Microwave

These are very expensive tags that are recycled by the storeowner, after removing them from the goods on sale. They are attached to the goods with special clips that can only be removed with the appropriate tool. Reading range of the scanning equipment is large and poorly defined.

The above four technologies are not interchangeable, and it is likely that different technologies will be favored in different parts of the world.

9.1.5.9 Magnetic Coupled Transponder Systems

These are the most common transponders available today, manufactured by a wide range of suppliers.

Generally operating at frequencies typically in the order of 125 KHz, the tags are characterized by antenna systems that comprise numerous turns of a fine wire around a coil former to collect energy from a reader's magnetic field. Due to the magnetic method of coupling, range is limited generally to a number of inches, being determined by the fields generated between the effective North and South Pole of the reader. Magnetic Tags are manufactured by many suppliers and find application in tagging animals, labeling gas bottles, electronic automobile key identification, and factory automation.

The different manufacturers use different forms of communication. Typical methods are to use the energizing signal at a frequency of 125 kHz and to receive data back from the transponder by:

- Receiving data back from the transponder at half the frequency of the transmitter link while the transmitter operates in a CW mode.

- Using the transmitter in pulse mode, and to transmit the data back immediately the transmitter signal stops, namely on the flyback while the tag energy is decaying when the energy is removed.
- By letting the tag load the energizing field, with this fluctuating load being sensed by the changes occurring on the energizing field.

Issues around magnetic coupling are that the frequency is low, the energizing field is very much stronger than the returned data field strength, that it is difficult to create filters with sufficient tuning to separate the transmit and received signal while both are present, that the tags have very limited energy storage capability, meaning that the energizing field needs to be applied in a uniform continuous manner, or data can be received back in that short period of time after the energizing field is removed (flyback).

Since 1998 a new series of tags that are actually magnetic coupled, but which operate on similar principles to their electric field coupled counterparts are becoming available. The tags are comprised of a small coil of a few turns, often etched on a flexible printed circuit substrate, and to which a single chip is bonded. These transponders might be as small as 1.5 cm by 1.5 cm in area and couple their energy to the reader via magnetic propagation. By the reader continuously providing an energizing field, which can be modulated, the tags can extract energy and data from the reader and communicate back to the reader. Such tags often have read/write capabilities and often anti-collision properties to allow for many tags to be in the reader beam at the same time. The reading and writing distance of such tags is limited by the magnetic means of propagation to typically 18 cm, but some manufacturers claim a 1-meter operating range. These tags seem to be positioning to replace the more difficult to produce 125 KHz tags which required coil winding facilities.

Almost all magnetic-based transponder systems are passive, which means that they get their energy from the reader's energizing field.

Transponder systems operating on magnetic coupling principles operate at frequencies as high as 29 MHz.

9.1.5.10 Electric Coupled Transponder Systems

Electric field coupled transponders generally provide vastly increased ranges over their magnetic counterparts. Rather than being limited to the ranges of the lines of force emitting from a magnetic field generator, they use the electric field propagation properties of radio communication to convey energy and data from the reader to the transponder and data from the transponder to the reader.

Electric field propagation requires antenna systems that are typically half a wavelength of the operating frequency in size (150 cm at 100 MHz, 15 cm at 1 GHz, 5 cm at 2.5 Ghz and 2.5 cm at 5.8 Ghz). This causes practical limits of how low a frequency should be used to start using Efield propagation methods, due to the size of the antenna.

Higher operating frequencies require more expensive components and lose the ability to transfer energy at a rate of the inverse of the wavelength squared.

In addition, the energy density of a signal radiated using electric field coupling decreases as the inverse of the distance squared between the source and the transponder. Whereas sensitive receivers can compensate for this loss of energy for the data communications over long distances, passive transponders which use the reader's energizing field as a source of power are practically limited to maybe 10 meters (say at 400 MHz). Beyond that distance (which reduces drastically with increased frequency to less than 1 meter at 2.5 GHz) it is necessary for the tags to use an external battery as a source of power.

Electric field tags are available in many different configurations and price ranges, particularly dependant on the complexity of the transponder. If the transponder is a read/write transponder and is required to operate beyond the range of passive transponders, the receiver circuitry onboard can be expensive and difficult to construct particularly if frequency stability is needed with temperature.

However, the invention of the backscatter modulation principle at Lawrence Livermore Laboratories in the 1960s and the skills of semiconductor designers to shrink all features into cheap integrated circuits, has meant that electric field-type tags in a read-only mode can be made extremely cheaply, most probably for less than 10 US cents in high volume. Such a tag would be passive, have no onboard tuned circuits, be read only, consist of a single integrated circuit and a simple antenna, would operate at any of a range of frequencies, be temperature insensitive, and would broadcast a large data value when illuminated by a reader's energizing field. In such a system the reader is complex because it provides the frequency stability, the energy of the system, and the receiver selectivity to receive the weak return communications, but the tags are very cheap. This is ideal for the situations where there is one reader and many tags.

Electric field tags need to operate in an ordered spectrum management system as their radiated energy (particularly from the reader) can be detected by other sensitive receivers far away and cause possible interference.

Recent developments in passive tag technology see the amount of power needed to power up the tag dropping dramatically. The reader radiates energy from its transmit antenna, some of which is collected by the tag in an area around its antenna called the 'antenna's aperture'. The size of this area is dependant upon the characteristics of the tag antenna and the operating frequency of the system, (e.g. a 915 MHz dipole has a 134 cm^2 aperture). Traditionally, a five-volt logic circuit in a transponder would need 55 milliwatts of RF energy to operate, while recent developments see this amount of power dropping to less than 1 milliwatt, thereby dramatically reducing the power needed by the reader and increasing the range over which passive transponders can operate effectively.

Recent developments with electric field tags relate to the development of transponder/smart card systems for toll road applications. Here the tags are active (that is they have a battery) but only consume battery power after the tag is 'activated' by passing through a high-energy activation field. Thereafter, the tag can send/receive data with an overhead reader and can adjust the data representing the balance remaining in the smart card after the toll fees are deducted. Such applications are proposed in the 2.45 GHz frequency band and more recently in the 5.8 GHz band.

A separate category also exists of 'active' tags (battery powered). These tags are 'beacon' tags, that is they are not interrogated by a reader, but wake themselves up from a low power 'sleep mode' periodically and broadcast their identity before returning to 'sleep mode'. By broadcasting on a fixed frequency, a sensitive receiver tuned to that frequency and within close proximity to the tag will receive the identity message. This type of transponder offers ranges up to hundreds of meters, but is not suited for situations where the location of a tag is being determined to a couple of meters range, or where very many tags are present in the reader zone. Encryption technology has also been added to these systems to stop unwanted tags being accepted as valid codes by the reader.

Despite the hurdles, the greater range, higher data rates and new technologies make these transponders suitable for a great number of applications.

9.1.6 How RFID Systems Work

9.1.6.1 A Brief Introduction to Some of the Systems Available

A radio transponder system uses radio energy to provide an energizing field, which powers up the transponders in the field and enables them to return their identity back to the reader. The transponders can be attached to goods and the radio energy can penetrate many types of packaging, allowing a system to, for example, scan the contents of a supermarket trolley, a pile of books, goods in a warehouse, etc. It is actually not as simple as this but we will attempt to provide further information to highlight some of the issues.

An Electronic Article Surveillance system (EAS) is used to detect the presence of operating tags that pass through its energizing field as shown in Figure 9.1. It could be used for instance in a retail outlet to detect tags that have not been

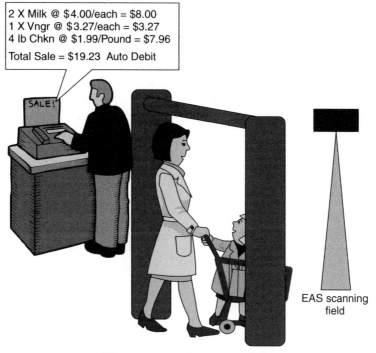

2 X Milk @ $4.00/each = $8.00
1 X Vngr @ $3.27/each = $3.27
4 lb Chkn @ $1.99/Pound = $7.96

Total Sale = $19.23 Auto Debit

EAS scanning
field

Electronic Article Surveillance (EAS)

Figure 9.1   EAS

neutralized at a point of sale terminal, therefore acting as an effective anti-shoplifting system.

The components of a tagging system are the transponders, a transmitter to provide energy to the tags, and a reader to receive the identities of the tags, and a computer system to process the information as shown in Figure 9.2. In order to receive the identities of many tags in a field at the same time, a protocol is needed. By using electric field propagation systems, typical reading ranges of 4 meters are easily achievable.

There are two major propagation methods that have an influence over the range of detection of the transponders from the readers. In magnetic systems, the energizing field is similar to that shown at school between the poles of a permanent magnet shown with iron filings. To increase the range the poles of the magnet need to be separated (increase the size of the reader) or increase the power of the field (not very effective). Coupling is by means of coils for the antennas, and typically the energizing frequency could be 130 KHz. Receiving antenna could have 1000 turns on the antenna.

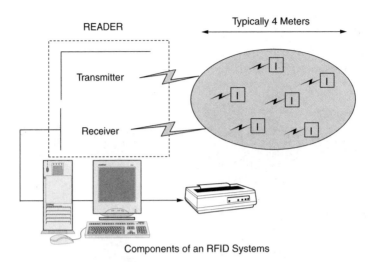

Components of an RFID Systems

Figure 9.2   RFID components

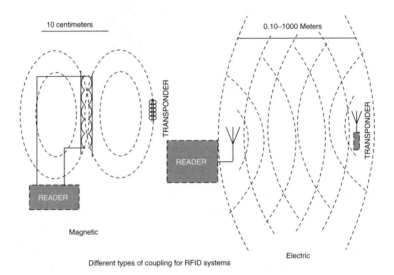

Different types of coupling for RFID systems

Figure 9.3   RFID coupling types

The electric field coupling uses antennas to propagate the energy from the transmitter to the transponder. The antennas' size is dependent on the frequency of operation (typically 400 MHz to 2.5 GHz) and the energizing field typically drops off as the inverse of the square of the distance between the transmitter and the transponder. Using this method, ranges of 4 meters are achievable. See Figure 9.3.

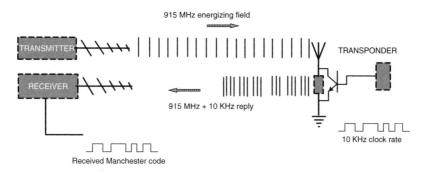

Figure 9.4   Backscatter modulation frequency shift

Some transponder systems can use a principle called *backscatter modulation* to communicate the data from the transponder to the reader, when in the presence of an energizing field. See Figure 9.4

The energizing field is emitted from the transmitter in the form of a carrier wave signal at a fixed frequency (in the example this is 915 MHz). This energy from the transmitter is collected by the transponder antenna, rectified and used to power the transponder. The transponder generates a data stream comprising a clock signal, and the data to be communicated is a form of a modified Manchester code. Typically the clock rate of the transponder might be at 10 KHz. The data from the transponder are used to drive a shorting transistor across the antenna, which has the effect of changing the reflectivity of the transponder antenna and causing some of the received energy from the transmitter to be reflected back towards the receiver. This reflected energy has the form of packets of energy at a frequency that is shifted from the original transmitter carrier by the clock rate of the transponder.

A simple receiver using the transmitter signal as a local oscillator can decode the received energy and extract the modified Manchester code.

10

Connecting the Last Mile

10.1 Introduction

Almost anything that can be done on a wired network can also be done on fixed-wireless broadband systems. This includes:

- T-1 Circuit
- Cable Television cable
- Ethernet Cable
- Fiber optic Cable.

Fixed-wireless broadband systems are designed so that they emulate cable connections and can use the same type of interfaces and protocols as T-1s, frame relay, Ethernet and ATM. Fixed-wireless systems are also used for voice communications and for carrying television programming. Fixed-wireless systems match the capabilities of cable-based systems for important parameters including delay, bit-error rate (1 in 100 million or better, and throughput (1 to 155 Mbps).

Any application that operates over a cable or wire-line circuit should be able to operate over a fixed-wireless system with the exception of the use of geosynchronous satellites where delays can exceed a quarter of a second.

There are times when a fixed-wireless connection is more practical and cost competitive with available wireline connections. Today fewer than 14% of buildings have fiber to them and only about 54% are close enough to a central office (12 000 feet or 3.5 km) to take advantage of DSL technology.

Wireless Broadband offers tremendous flexibility and improving performance but it does have some limitations. Wireless broadband uses radio spectrum, which

Wireless Data Technologies. Vern A. Dubendorf
© 2003 John Wiley & Sons, Ltd ISBN: 0-470-84949-5

Table 10.1 The principle frequencies used for fixed wireless broadband

900 MHz, 2.4 GHz, and 5.8 GHz	Unlicensed systems using spread-spectrum technologies.
2.5 GHz	Licensed to carriers for MMDS (Multichannel Multipoint Distribution System).
5 GHz	Unlicensed band referred to as the UNII (Unlicensed National Information Infrastructure) band.
23 GHz	Commonly used for microwave LAN systems.
28 GHz	Licensed to carriers for LMDS (Local Multipoint Distribution Service).
38/39 GHz	Licensed to carriers for general-purpose communications services.

is a limited resource. This limits the number of wireless broadband users and the amount of spectrum available to any user at any moment in time. The amount of spectrum available equates almost directly to data bandwidth, 1 Hz of spectrum typically yields between 1 and 4 bps of throughput depending on factors such as modulation type and environment. The amount of spectrum available varies by radio band. The frequencies used for Wireless Broadband most often require 'line-of-sight' and are limited to distances that vary from a few miles to tens of miles. Fiber optic cable offers the greatest overall capacity. Table 10.1 provides a list of the principle frequencies used for fixed wireless broadband.

The frequency allocation for fixed wireless ranges from the 900 MHz to 40 GHz ranges. Higher frequencies provide for more spectrums available for broadband applications, the majority of which are found above 10 GHz. In these frequency ranges, antennas are normally smaller due to the smaller wavelengths, which makes deployment easier. There are some drawbacks however. Because of the higher frequencies, components require more complex technology, which in turn increases cost. In addition, propagation distances for reliable communications decreases and the signal is more susceptible to weather conditions such as rain, fog, and snow. Systems with frequencies above 30 GHz are referred to as millimeter wave systems because the wavelength of those signals are on the order of 1 mm in length.

10.2 LMDS

Licensed point-to-point connections found in the frequency range from 1.7 to 40 GHz. The lower frequencies are most commonly used by carriers for backhaul networks such as T3 connectivity with a 45 Mbps throughput. Multi-hop systems normally operate at 2, 4 and 6 GHz with a normal high-end throughput of 155 Mbps (OC-3) but there is no inherent upper limit. The FCC has set aside a frequency range of 21.2 to 23.6 GHz for private use. Specifically, telecommunication companies for backhaul networks such as PCS use the rest of the range.

Licensed frequencies virtually eliminate the possibility of interference and are pretty reliable and provide considerable head room for increasing throughput in case your needs expand in the future.

In order for private companies to install point-to-point licensed connections, the FCC has streamlined the license application process. The average cost for a fully installed 10 Mbps connection with a 5-mile (8-km) range runs to about US$30 000.00.

11

Wireless Information Security (W-INFOSEC)

We have touched on Wireless Security throughout this book. In this chapter it is my desire to go deeper into the subject of Information Security in general as well as its impact on Wireless Technologies.

11.1 Introduction

In this day and age, almost every major newspaper, or network news broadcast, and in every Information Technology-oriented news bulletin and magazine, there are headlines about computer security break-ins and system vulnerabilities. These are providing the drive behind launching information security, or InfoSec as it is most often referred to, programs.

Times have changed and Information Security programs now need to confront much more than basic security subjects. Many experts believe that it is important to develop a security program within the context of the company's business objectives and culture in order to better understand where the risk comes from and why. Therefore, an InfoSec program needs to uncover and respond to the business risks present in the organization. This process begins by critically assessing aspects of the practices and procedures from the standpoint of security. This can reveal unalterable aspects of the corporate culture, initiatives with a higher priority than security, staff and operational limitations, retention and recruitment difficulties, budget constraints, etc.

Since its inception, Information Security has been somewhat of a dark horse in Computer Science. The past few decades have seen a global drive to realizing

Wireless Data Technologies. Vern A. Dubendorf
© 2003 John Wiley & Sons, Ltd ISBN: 0-470-84949-5

the potential of the information revolution in a desire to make a positive change on the world. Now that we are deep inside the information age, we are now faced with addressing the negative manifestations of these advancements in an environment that was designed to be open – not protected. As has often occurred in the past, attention to information security is being mandated by consequences. The development of the Internet as a global network for the instantaneous distribution of knowledge, products and ideas, and its availability to the general public, has enabled hackers and crackers to commit crimes around the world with little chance of repercussion and given rise to the new phenomenon of identity theft. In conjunction with the free dissemination of security vulnerability information and tools, the Internet has created a worst-case situation for protecting our systems.

Although incorrect, to many managers and users the security of data on wired network is an assumed fact. They do not however, extend this assumption to wireless data. They instead show a high level of discomfort. This is something that wireless specialists find strange. It is an unpleasant fact that any network, whether using wire, fiber, or air, is subject to security risks. These include:

- Threats to the *physical* security of a network;
- Attacks from within the network's (authorized) user community;
- Unauthorized access and eavesdropping.

It is a fact that wireless LANs do indeed maintain the exact same properties of wired LANs, but it is done without the need for copper or fiber, and the steps needed to maintain the security and integrity of data applies to both environments. All network services, including their pitfalls, remain the same with the exception of the physical layer and SHY, which is the only real difference between being wired or wireless. Wireless LAN technology actually includes a set of security elements that is not available in a wired network. This is the main point, wireless specialist believe, that makes the wireless network even more secure than copper- or fiber-based networks, and this opinion is shared by an overwhelming majority of industry analysts and experts.

11.2 Public Key Infrastructure (PKI)

Networks are increasingly being used as the backbone of mission-critical electronic transactions and commerce. The upside is that this provides instant access to the people who need the information. The downside is that it opens up key corporate systems to potential security risks.

Consequently, trusted and affordable network security is a must, offering many opportunities to establish important competitive advantages and improve business processes. Increasingly, organizations rely on secure e-mail, electronic forms, intranets, extranets, and virtual private networks (VPNs) to maximize their effectiveness in a competitive global economy. In today's global e-business environment companies need to trust their business partners and know that their information is kept private. For higher dollar or volume transactions, customers need to know that the transaction is legally binding and the content is delivered unchanged.

Some industry analysts suggest that the reasons why PKI isn't more widely used are a lack of planning, cost of PKI deployment or lack of internal communication of business value.

It is not that PKIs are technically too complicated. On the contrary, a PKI is a fairly simple installation that can be up and running in a lab within a few hours. But the business uses of a PKI are harder to grasp. Many think it is an elixir, a solution to all that ails a company.

11.3 What is a PKI?

Public-key cryptography provides the foundation of network security through encryption and digital signatures. Together, encryption and digital signatures provide:

- Authentication. Allows your e-business to engage trusted customers, partners and employees.
- Authorization. Allows business rules to dictate who can use what resources, under what conditions.
- Confidentiality. Protects confidentiality of sensitive information, while stored or in transit.
- Integrity. Prevents any transaction from being tampered with and will notify you not to trust the contents should the message change from its original state.
- Non-repudiation. Prevents any party from denying an e-business transaction after the fact.
- Audit controls. Provides audit trails and a record of critical and non-critical events that have occurred within the Entrust infrastructure.

All of these security benefits are essential to conduct truly secure electronic business transactions.

11.4 PKI and Other Security Methods

There are many different security methods being used today. High levels of security do not have to be used for every application across the board. For instance, some internal applications such as e-mail require normal security whereas other internal applications such as payroll or legal contracts require a higher level of security. Each company needs to assess which type of security methods fit their business requirements.

11.4.1 Username/Password

This is the easiest to implement and most likely everyone has one anyway. There is no special software required and it is easily scalable. However, it is one of the lowest types of security and does not provide Digital Signatures. Many IT professionals find it expensive to manage with respect to password changes and resets. In addition, too many passwords for various applications cause user confusion.

11.4.2 Biometrics

This is the most guaranteed type of authentication. It cannot be forged and is easy to use and remember. As a result, it is very expensive to implement and manage and is not generally scalable.

11.4.3 Tokens/Smart Cards

Tokens and smart cards provide a stronger password authentication because of random generation of passwords. This makes is easier for the user because they don't need to remember a specific password. However, because it is tangible, tokens can be easily lost or stolen. This causes IT departments to incur a higher expense since tokens can often cost $100–$200 a piece. In addition, smart cards often create interoperability issues since not all applications can use them.

11.4.4 SSL Protected Messages

SSL is not generally thought of as a message security method since it has limits. SSL can be used in customer service messaging applications whereby data is entered on a Web form or in a free-text message application. While the information is protected in the transport channel, it is not protected after it passes through the server on its way to the application area. Accordingly, SSL does not provide 'end-to-end' security. SSL also provides assurances that the Web server belongs to the assumed party. (SSL does not validate the customer's identity, but the enterprise can do so by other mechanisms such as PKI.)

Organizations in every field and industry use PKI to build relationships founded on trust with employees, partners, and customers.

11.4.4.1 Corporations

With a PKI in place, companies can use digital certificates to replace easily forgotten and cracked user IDs and passwords, enabling secure 'single login'. Employees can safely access everything from HR enrollment forms to 401(k) data, and take advantage of e-mail authenticated and signed by digital certificates.

11.4.4.2 Financial Services

Banks and brokerage houses implement PKIs to give customers secure access to account information, allowing them to initiate trades and transfer funds with confidence.

11.4.4.3 Health Care Organizations

HMOs let customers securely check claim status and submit data without fear that private information will be intercepted or corrupted.

11.4.4.4 Software Distributors

Software companies with a PKI 'digitally shrink-wrap' software downloaded via the web to customers, who know the software is genuine and has not been tampered with.

11.4.4.5 Publishers

Magazines and news organizations deliver content online using a PKI to verify customer identities, grant access to different subscription levels, and assure readers that the content is coming from the authentic source, not a 'spoofer'.

Before organizations can begin implementing PKI and acting as a Certificate Authority (CA) in issuing certificates, they need to be able to issue certificates that contain company-specific identifying information, and they must be able to control who is issued a certificate. There are two main PKI options:

- Closed PKI. With proprietary PKI software, digital certificates are issued to a limited, controlled community of users. Applications – including those of extranet users and anyone else outside the enterprise with which employees need to communicate securely – need a special software interface from the PKI vendor to work with the certificates. Closed PKI systems require additional training, hardware, software, and maintenance.
- Open PKI. Applications interface seamlessly with certificates issued under an open PKI, the roots of which are already embedded. Open PKI systems allow enterprises to become their own CA, while taking advantage of the PKI vendor's service and support.

11.5 Digital Certificates

Digital certificates are electronic files that are used to uniquely identify people and resources over networks such as the Internet. Digital certificates also enable secure, confidential communication between two parties.

When you travel to another country, your passport provides a universal way to establish your identity and gain entry. Digital certificates provide similar identification in the electronic world. Certificates are issued by a trusted third party called a Certification Authority (CA).

Much like the role of the passport office, the role of the CA is to validate the certificate holders' identity and to 'sign' the certificate so that it cannot be forged or tampered with. Once a CA has signed a certificate, the holder can present their certificate to people, web sites, and network resources to prove their identity and establish encrypted, confidential communications.

A certificate typically includes a variety of information pertaining to its owner and to the CA that issued it, such as:

- The name of the holder and other identification information required to uniquely identify the holder, such as the URL of the web server using the certificate, or an individual's e-mail address.
- The holder's public key. The public key can be used to encrypt sensitive information for the certificate holder.
- The name of the Certification Authority that issued the certificate.

- A serial number.
- The validity period (or lifetime) of the certificate (a start and an end date).

In creating the certificate, this information is digitally signed by the issuing CA. The CA's signature on the certificate is like a tamper-detection seal on a bottle of pills – any tampering with the contents is easily detected.

Digital certificates are based on public-key cryptography, which uses a pair of keys for encryption and decryption. With public-key cryptography, keys work in pairs of matched 'public' and 'private' keys. In cryptographic systems, the term key refers to a numerical value used by an algorithm to alter information, making that information secure and visible only to individuals who have the corresponding key.

The public key can be freely distributed without compromising the private key, which must be kept secret by its owner. Since these keys only work as a pair, an operation (for example encryption) done with the public key can only be undone (decrypted) with the corresponding private key, and vice-versa.

A digital certificate securely binds your identity, as verified by a trusted third party (a CA), with your public key.

A WAP server WTLS certificate is a certificate that authenticates the identity of a WAP site to visiting micro-browsers found in many mobile phones on the market.

11.6 Wireless Transport Layer Security (WTLS)

11.6.1 WTLS

A huge growth of the wireless mobile services poses demand for the end-to-end secure connections. The Wireless Transport Layer Security provides authentication, privacy, and integrity for the Wireless Application Protocol. The WTLS layer operates above the transport protocol layer. It is based on the widely used TLS v1.0. The requirements of the mobile networks have been taken into account when designing the WTLS; low bandwidth, data gram connection, limited processing power and memory capacity, and cryptography exporting restrictions have all been considered.

11.6.1.1 WTLS Class 2

WTLS Class 2 provides the capability for the client to authenticate the identity of the gateway it is communicating with. Table 11.1 gives an overview of the steps necessary to enable WTLS class 2.

Table 11.1 Steps to enable WTLS class 2

Two Phase Security Model

(1) The Gateway sends a certificate request to the PKI Portal.
(2) The PKI portal confirms the ID and forwards request to the CA.
(3) The CA sends the Gateway Public Certificate to the Gateway (may be via Portal).
(4) WTLS Session established between Phone and Gateway.
(5) SSL/TLS session established between Gateway and Server.

Future Additions to Provide an 'End-to-End Security Model'

(6) Server sends certificate request to PKI Portal.
(7) Portal Confirms ID and forwards request to CA.
(8) CA sends Server Public Certificate to Server.
(9) WTLS Session Established from Phone to Server (routing is via Gateway, but communication is opaque to Gateway).

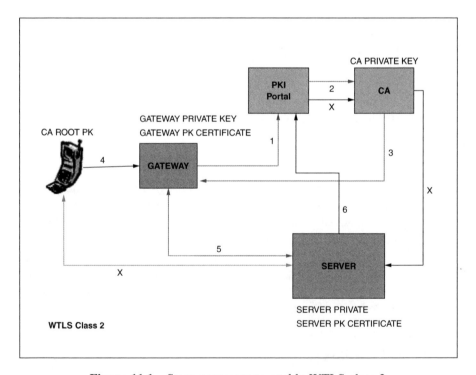

Figure 11.1 Steps necessary to enable WTLS class 2

In Table 11.1 the device is provisioned with some CA Root Public Key information. The WAP Gateway generates a key pair (public key and private key).

Currently WTLS operates between a WAP client and a gateway and future versions of WAP will allow a WTLS session to terminate beyond the gateway at an application or origin server. See Figure 11.1.

11.6.2 WAP

The Wireless Application Protocol (WAP) is a result of continuous work to define an industry-wide specification for developing applications that operate over wireless communication networks.

WAP security functionality includes the Wireless Transport Layer Security (WTLS) and application-level security, accessible using the Wireless Markup Language Script (WMLScript). The security provided in WAP can be of various levels. In the most basic case, anonymous key exchange is used for creation of an encrypted channel between server and client. The next level of security is where a server provides a certificate mapping back to an entity trusted by the client. Finally the client may possess a private key and public key certificate enabling it to identify itself with other entities in the network.

The WAP Identity Module (WIM) is used in performing WTLS and application-level security functions, and especially, to store and process information needed for user identification and authentication. The WPKI may use the WIM for secure storage of certificates and keys.

11.6.3 WEP

There is a huge amount of information in the IEEE 802.11 standard and its extensions. IEEE standards are divided into clauses and annexes. Information in the standard is referred to by the clause and the annex in which it is found.

Clause 7 of the standard describes the MAC frames and their content.

Clause 8 of the standard describes the WEP functionality that may be implemented in an IEEE 802.11 station.

The IEEE 802.11 standard incorporates MAC-level privacy mechanisms to protect the content of data frames from eavesdropping. This is due to the fact that the medium to be used by the IEEE 802.11 Wireless LAN (WLAN) is extremely different from that of the wired LAN. The WLAN does not have the minimal privacy provided by a wired LAN. The wired LAN must be physically attacked or compromised in order to tap into its data. A WLAN, on the other hand, can be attacked or compromised by anyone with the proper type of antenna. The IEEE 802.11 Wired Equivalent Privacy (WEP) mechanism provides protection at a level that is felt to be equivalent to that of a wired LAN, hence its name Wired Equivalent Privacy.

WEP is an encryption mechanism that takes the content of a data frame, its frame body, and passes it through an encryption algorithm. The result then replaces the frame body of the data frame and is transmitted. Data frames that are encrypted are sent with the WEP bit in the frame control field of the MAC header set. The receiver of an encrypted data frame passes the encrypted frame body through the same encryption algorithm used by the sending station. The result is the original, unencrypted frame body. The receiver then passes the unencrypted result up to higher layer protocols.

We need to be aware of the fact that only the frame body of data frames is encrypted. This leaves the complete MAC header of the data frame, and the entire frame of other frame types unencrypted and available to eavesdropping. Thus, WEP does provide protection for the content of the data frames, but we really need to be aware of the fact that it does not protect against other security threats to a LAN, such as traffic analysis.

The encryption algorithm used in IEEE 802.11 is RC4. Ron Rivest of RSA Data Security, Inc. (RSADSI), developed RC4. RSADSI is now part of Network Associates, Inc.

RC4 is a symmetric stream cipher that supports a variable length key. A symmetric cipher is one that uses the same key and algorithm for both encryption and decryption. A stream cipher is an algorithm that can process an arbitrary number of bytes. This is contrasted with a block cipher that processes a fixed number of bytes.

The key is the one piece of information that must be shared by both the encrypting and decrypting stations. It is the key that allows every station to use the same algorithm, however only those stations sharing the same key can correctly decrypt encrypted frames. RC4 allows the key length to be variable up to 256 bytes as opposed to requiring the key to be fixed at a certain length. The IEEE 802.11 has chosen to use a 40-bit key.

The IEEE 802.11 standard describes the use of the RC4 algorithm and the key in WEP, although key distribution or key negotiation is not mentioned in the IEEE 802.11 standard. This leaves much of the most difficult part of secure communications to the individual manufacturers of IEEE 802.11 equipment. In a secure communication system using a symmetric algorithm, such as RC4 it is extremely important and necessary that the keys used by the algorithm be protected and secrecy maintained. If a key is compromised all frames that have been encrypted with that key are also compromised. Therefore, while it is likely that equipment from the majority of the manufacturers will be able to interoperate and exchange encrypted frames, it is highly unlikely that a single mechanism will be available that will securely place the keys in the individual stations. There is currently discussion in the IEEE 802.11 working group to address this lack of standardization.

The IEEE 802.11 provides two mechanisms to select a key for use when encrypting or decrypting a frame.

The first mechanism is a set of up to four default keys. Default keys are intended to be shared by all stations in a BSS or an ESS. The benefit of using a default key is that once the station obtains the default keys a station can communicate securely with all of the other stations in a BSS or ESS. The problem with using default keys is that they are widely distributed to many stations and may be more likely to be revealed.

The second mechanism provided by IEEE 802.11 allows a station to establish a 'key mapping' relationship with another station. Key mapping allows a station to create a key that is used with only one other station. Although this one-to-one mapping is not a requirement of the standard, this would be the most secure way for a station to operate, since there would be only one other station that would have knowledge of each key used. The fewer the stations having possession of a key, the less likely the key will be revealed.

The dot11PrivacyInvoked attribute controls the use of WEP in a station. If dot11PrivacyInvoked is false, all frames are sent without encryption. If dot11PrivacyInvoked is true, all frames will be sent with encryption, unless encryption is disabled for specific destinations. Encryption for specific destinations may only be disabled if a key mapping relationship exists with that destination.

A default key may be used to encrypt a frame only when a key mapping relationship does not exist between the sending and receiving station. When a frame is to be sent using a default key, the station determines if any default keys are available. There are four possible default keys that might be available. A key is available if its entry in the dot11WEPDefaultKeysTable is not null. If one or more default keys are available, the station chooses one key, by an algorithm not defined in the standard, and uses it to encrypt the frame body of the frame to be sent. The WEP header and trailer are appended to the encrypted frame body, the default key used to encrypt the frame is indicated in the KeyID of the header portion along with the initialization vector, and the integrity check value (ICV) in the trailer. If there are no available default keys, i.e., all default keys are null the frame is discarded.

If a key mapping relationship exists between the source and destination stations, the key-mapping key, the key shared only by the source and destination stations, must be used to encrypt frames sent to that destination. When a frame is to be sent using a key-mapping key, the key corresponding to the destination of the frame is chosen from the dot11WEPKeyMappingsTable, if the dot11WEPKeyMappingWEPOn entry for the destination is true. The frame body is encrypted using the key-mapping key, and the WEP header and trailer are

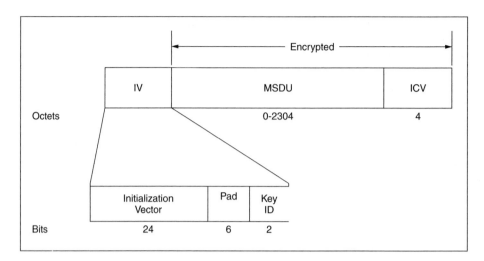

Figure 11.2   WEP expansion of the frame body

appended to the encrypted frame body. See Figure 11.2 for the WEP expansion
of the Frame Body. The value of the KeyID is set to zero when a key-mapping
key is used. If the value of dot11WEPKeyMappingWEPOn for the destination is
false, the frame is sent without encryption.

Corresponding to the dot11PrivacyInvoked attribute controlling the sending
of frames, the dot11ExcludeUnecrypted attribute controls the reception of
encrypted frames. When dot11ExcludeUnecrypted is false, all frames addressed
to the station are received, whether they are encrypted or not. However, when
dot11ExcludeUnecrypted is true, the station will receive only frames that are
encrypted, discarding all data frames that are not encrypted. If a frame is discarded
because it is not encrypted and dot11ExcludeUnecrypted is true, there is no
indication to the higher layer protocols that any frame was received.

There are two counters associated with WEP. The dot11UndecryptableCount
reflects the number of encrypted frames that were received by the station that could
not be decrypted, either because a corresponding key did not exist or because the
WEP option is not implemented. The dot11ICVErrorCount reflects the number of
frames that were received by a station for which a key was found that should have
decrypted the frame, but that resulted in the calculated ICV value not matching the
ICV received with the frame. These two counters should be monitored carefully
when WEP is used in a WLAN. The dot11UndecryptableCount can indicated that
an attack to deny service may be in progress, if the counter is increasing rapidly.
The dot11ICVErrorCount can indicate that an attack to determine a key is in
progress, if this counter is increasing rapidly.

11.6.4 WPKI

The goal of the WAP PKI is to reuse existing PKI standards where available, and only to develop new standards where it is necessary to support the specific requirements of WAP. To the extent possible, the WAP PKI will work interchangeably with existing X.509v3 certificates in existing Internet applications in order to leverage the existing Internet PKIs.

The current WAP PKI model defines the functionality that is needed to manage the security functionality defined in WAP1.2 that is summarized as:

- CA Public Key Certificates used for WTLS Class 2.
- Client Public Key Certificates used for WTLS Class 3.
- Client Public Key Certificates used in conjunction with WMLScript signText.

WAP PKI Model versions in the future will be enhanced to provide support for signed content models for protecting the download of WMLScript and WTA Scripts to the client as well as functionality support for the application level end-to-end confidentiality and integrity. Currently, it is assumed that clients will not possess keys to allow end-to-end confidentiality and they will not support verification of signatures on scripts.

The following is the general model adopted in the current version of the WPKI:

- WTLS Server and Root CA certificates stored in the device will be according to WTLSCertificate defined in (WAPWTLS).
- Client certificates (WTLS and application) and CA Roots stored in servers will be according to X.509 as profiled in (RFC2459).
- Client certificates (WTLS and application) and CA Roots that are to be sent OTA and/or stored in WAP client devices will be according to X.509 as profiled in (CERTPROF).
- Storage of the certificate URL in the device, rather than the full client certificate, is the preferred model, when X.509 format certificates would otherwise be expected to be transferred OTA.
- Storage of X.509 client certificates in the device is expected to be the exception, unless they are provisioned on the device, through the (WIM) for example.

Certificates can be stored in several locations on the client, on a WIM (either on the same ICC as the SIM or not in a client that supports a SIM) or on the client itself.

11.6.4.1 Private Key Capability

It is assumed that the WAP clients are able to have at least two different signing keys:

- One for WTLS client authentication, and
- One for application layer signing (signText).

It is planned that the majority of all clients will fit into one of the following classes:

- No private keys
- One private key (either for authentication or signing)
- Two private keys (one for authentication and one for signing).

Whether for authentication or signing, in order for a WAP client's signature to be trusted by servers there are times when it becomes necessary for the client to be registered in a new PKI that is trusted by the server. This shows the need for an application layer PKI registration functionality. A client that registers in such a PKI does not necessarily need to trust that PKI, so that installation of trusted CA information may not be required. However, the client may need to be able to trust the PKI in order to authenticate a WTLS server. The client is required to be able to identify itself in the server's trusted PKI.

The term PKI portal is used for any PKI entity (RA, CA or OCSP (RFC2560) responder) that the WAP client communicates with during PKI operations. There is no assumption that the client communicates directly with the PKI portal, though any security mechanisms applied (e.g. signature) must be end-to-end between the client and PKI portal. The PKI portal need not be co-located with either the service provider or operator.

In the majority of cases the registration of WAP clients in a PKI will have occurred as part of the provisioning of a WAP device. This specification also provides for a mechanism that allows this registration to be carried out, over the air, after the device has already been provisioned. A 'typical' PKI registration may involve the following types of interchanges:

- The client contacts a service provider such as a content provider that supplies some type of health care application, attempting to use a service that requires a client signature.
- The Service provider requires the client to be registered in its chosen PKI. The Service provider indicates to the client:

- That the client should contact a PKI portal. It may also provide some PKI information such as a CA name.
- The client contacts a PKI portal and submits certification request. The PKI portal acknowledges the receipt of request. The acknowledgment message gives guidance to the client as to what will happen next.
- The Normal certification processes occur in the PKI. This may result in near instant certification or may involve a significant time lapse.
- At some later time the client reconnects to the service provider. When producing a signature the client also includes information to identify its certificate.
- The service provider may use this to retrieve the client's certificate from a repository and can then verify the client's signature.

11.7 Authentication and Integrity

According to Schneier, authentication is defined as follows. 'It should be possible for the receiver of a message to ascertain its origin; an intruder should not be able to masquerade as someone else.' (Schneier, 1996).

Authentication is a technique to ensure that the stated identity of the user is correct. In the beginning, the other party introduces itself and claims to have some identity. This is not enough. The contacted party also needs to know for sure that the contacting party is the one it claims to be. The contacting party has to present some verification to prove its identity. It can be as simple as passwords, or more complicated digital signatures or certificates. But then again, the contacting party also wants to be sure that the other end is valid. The contacted party has to present some identification about itself.

After the authentication, the service provider can be sure that the service is available to the user who has correct rights to use the service. On the other hand, the user can be confident about the service provider.

Maintaining integrity means securing the reliability of the information. We have to figure out a way to prevent unauthorized changes or at least find the means to notice those modifications. Integrity is guaranteed by calculating checksums from the original information to be sent. Of course, just a plain checksum is not enough. We need some sender-related information mixed into calculations, e.g. information is signed with the user's digital signature.

In most cases, maintaining integrity is more critical than guaranteeing privacy. It is more important that the information is received unaltered, but checked by someone to ensure that nobody has been able to modify it, without making out the whole information. For example, bank transactions apply this category. It is embarrassing if someone finds out how much money you have but it is infuriating if somebody steals your money.

11.8 Security Threats

In this section we will look at specific security threats as well as look at some protective measures that can be taken. The threats include denial-of-service attacks, replay attacks, theft of information or passive eavesdropping, and session-stealing (for theft of information) attack.

11.8.1 Denial-of-Service Attack

A denial-of-service (DOS) attack is something done to preclude someone from accomplishing useful work. Typically, a DOS attack takes one of two forms: nuisance packets (TCP SYN flooding), or the preclusion of packets from flowing between two nodes. There is little that can be done to prevent this nuisance packet attack, and the sender can always spoof the source address. However, service providers can filter IP packets in their routers to assure the IP source address of a packet is genuine before it is forwarded. The consequence is that the starting point of the attack may be traced more accurately with this ingress filtering.

To perform the preclusion of packets flowing between two nodes, the attacker must be on the path between two nodes. If the attacker was to create a bogus Registration Request, stipulating his own IP address as the care-of address for a mobile node, the mobile node's home agent to the attacker would tunnel all packets. However, if cryptographically resilient authentication is compulsory by a mobile node and its home agent, there would be no difficulty. Mobile IP enables a mobile node to use the authentication algorithm of their preference. However, all must sustain the default algorithm of KEYED MD5. This authentication method draws on RFC 1321 to provide secret-key authentication and integrity checking.

11.8.2 Replay Attacks

It is feasible for an attacker to obtain a copy of a legitimate Registration Request, store it, and then replay it later to accomplish a forged care-of address for a mobile node. To avoid this replay attack from occurring, the mobile node produces a unique value for the identification field in each of the successive endeavors for registration. The identification field is made in such a way as to allow the home agent to ascertain what the subsequent value should be.

The attacker is hampered because the identification field in his stored Registration Request will be known as being outdated by the home agent.

11.8.3 Theft of Information or Passive Eavesdropping

This type of attack is against the confidentiality of the information. Encryption is the most common means used to protect data from unauthorized persons. There are at least two ways that data can be protected through encryption.

End-to-end encryption is the most thorough way to protect the data. This means encrypting and decrypting the data at the source and destination, as opposed to encrypting/decrypting over the first or last link. Some examples of Internet-based applications that provide such end-to-end protection include Secure Remote File Copy (SCP), Secure Sockets Layer (SSL), and Secure Remote Shell (SSH). The Encapsulating Security Payload RFC (1827) affords end-to-end encryption for other application programs that do not provide encryption themselves.

Link-layer encryption is classically used between a mobile node and its foreign agent of a wireless link. In this case, the mobile node and the foreign agent encrypt all packets they trade over the foreign link. Link encryption is particularly significant when the foreign link is a wireless LAN. It is easier to snoop a wireless link because no physical connection is required. RFC 1984 has more information on this topic.

11.8.4 Session-Stealing (for Theft of Information) Attack

An attacker performs a session stealing attack by waiting for a valid node to authenticate itself and initiate an application session, then captures the session by masquerading as the legitimate node. Typically, this requires the attacker to transmit numerous nuisance packets to thwart the legitimate node from recognizing that the session has been captured. This type of attack is disallowed by the above two methods of encryption: end-to-end and link-layer.

11.8.5 Secure Tunneling

Secure tunneling uses a firewall of the applications-layer type that also employs a cryptographically secure method for users to gain access to a private network across a public network. Both the IP Authentication Header and the IP Encapsulating Security Payload should be used.

Secure Tunnelers can also be used to create Virtual Private Networks (VPN) across a public network such as the Internet. A VPN behaves as a single, secure, logical network while being made up of numerous physical networks of varying levels of trust. The secure tunnel can shield private networks from being accessed by trespassers while providing confidentiality, which keeps a trespasser from eavesdropping on data exchanged between two networks.

11.9 HIPAA (USA)

The healthcare industry is facing a growing number of challenges with respect to regulations surrounding the confidentiality, integrity and availability of individual health information. This increasingly complex regulatory environment received momentum back on August 12, 1998 with the Notice of the Proposed Rule from the Department of Health and Human Services.

The Proposed Rule falls under the umbrella of the Health Insurance Portability and Accountability Act (HIPAA) that was passed on August 21, 1996. HIPAA contains a section entitled *Administrative Simplification* that the Health Care Financing Administration (HCFA) is responsible for implementing. On August 12, 1998 the HCFA and the Department of Health and Human Services released a Notice of the Proposed Rule concerning Security and Electronic Signature Standards (45 CFR, Part 142). This Proposed Rule suggests standards for the security of individual health information and electronic signature use for health plans, healthcare clearinghouses and healthcare providers. The health plans, healthcare clearinghouses and healthcare providers will use the Security Standards to develop and maintain the security of all electronic health information.

The Proposed Rule is not to be confused with Privacy legislation, which attempts to establish privilege rights for individual health information. The Security and Electronic Signature Standard attempts to establish the technical measures that guard against inappropriate access to individual health information.

12

Convergence: 3RD Generation Technologies

Third Generation (3G) is a generic name for a set of mobile technologies set to be launched by the end of 2001 which use a host of high-tech infrastructure networks, handsets, base stations, switches and other equipment to allow mobiles to offer high-speed Internet access, data, video and CD-quality music services.

Data speeds in 3G networks should show speeds of to up to 2 Megabits per second, an increase on current technology.

12.1 CDMA2000

The Telecommunications Industry Association (TIA) has adopted a specification based on Qualcomm's High Data Rate (HDR) which is considered to be a cost effective, high-speed, high-capacity wireless technology.

The HDR system is optimized for packet data services and has a flexible architecture based on IP protocols. HDR can overlay an existing wireless network or work as a stand-alone system. HDR unleashes Internet access by providing up to 2.4 Mbps in a standard bandwidth 1.25 MHz channel that is unprecedented in systems capable of fixed, portable and mobile services.

HDR, known as TIA/EIA/IS-856 'CDMA2000, High Rate Packet Data Air Interface Specification' is also known as 1xEV.

The 1xEV specification was developed by the Third Generation Partnership Project 2 (3GPP2), a partnership consisting of five telecommunications standards bodies: CWTS in China, ARIB and TTC in Japan, TTA in Korea and TIA in North America.

Wireless Data Technologies. Vern A. Dubendorf
© 2003 John Wiley & Sons, Ltd ISBN: 0-470-84949-5

The 1xEV technology is so versatile it can be embedded in handsets, laptops, notebooks and other fixed, portable and mobile devices. It can support e-mail, web browsing, e-commerce, telematics and many other applications. Imagine using your laptop to download office files while waiting between planes.

CDMA2000 offers:

- Air link provides up to 2.4 Mbps in a dedicated 1.25 MHz channel.
- Packet data design results in greatest spectral efficiency Flexible, IP-based architecture enables multiple implementation methods.
- Compatible with existing CDMA Networks.

12.2 CDMA2000 Types

There are various types of CDMA2000 types, as follows.

12.2.1 CDMA2000 1X

The 1xEV specification was developed by the Third Generation Partnership Project 2 (3GPP2), a partnership consisting of five telecommunications standards bodies: CWTS in China, ARIB and TTC in Japan, TTA in Korea and TIA in North America. It is also known as High Rate Packet Data Air Interface Specification. It delivers 3G-like services up to 140 kbps peak rate while occupying a very small amount of spectrum (1.25 MHz per carrier), protecting this precious resource for operators.

12.2.2 CDMA2000 1X EV-DO

1X EV-DO, also called 1X-EV Phase One, is an enhancement that puts voice and data on separate channels in order to provide data delivery at 2.4 Mbit/s. It was developed by the Third Generation Partnership Project 2 (3GPP2), a partnership consisting of five telecommunications standards bodies. Also known as High Rate Packet Data Air Interface.

12.2.3 CDMA2000 1X EV-DV

EV-DV, or 1X-EV Phase Two with promises of data speeds ranging from 3 Mbps to 5 Mbps. As many as eight proposals have been submitted to standards committee 3GPP2 for the design of EV-DV.

12.2.4 CDMA2000 3X

CDMA2000 3X is an ITU-approved, IMT-2000 (3G) standard. It is part of what the ITU has termed IMT-2000 CDMA MC. It uses 5 Mhz spectrum (3 × 1.25 Mhz channels) to give speeds of around 2–4 Mbps.

12.3 Operator Benefits of CDMA2000

1xEV is so flexible, it can be deployed as a stand-alone system, side-by-side with an existing or future voice system, or integrated into a current CDMA voice system. And because it installs easily using off-the-shelf retail components, consumers can set up the technology themselves, allowing operators to reduce costly, on-site service.

1xEV is the technology that helps you deliver high-performing, more cost-effective wireless data services to customers around the world. The 1xEV technology is compatible with CDMA voice systems and allows for side-by-side deployment to complement existing cellular/PCS networks. Because 1xEV can share cell sites, towers and antennas of these networks, you'll be able to deploy it more rapidly and get more out of the system you already have.

12.3.1 Air Link

HDR's highly efficient Air Link design achieves the type of data rates and performance levels you would only expect to see in next-generation technologies. Whether you are using fixed, portable or mobile applications.

12.3.2 Optimized Throughput

- Use of a single 1.25 MHz channel optimized for packet data results in greatest spectral efficiency.
- Peak data rate of 2.4 Mbps on the forward link and 307 kbps on the reverse link provides unprecedented speed.
- Average throughput on a loaded sector is an estimated 600 kbps on the forward link and 220 kbps on the reverse link.
- Dynamically assigned data rate adjusts as rapidly as every 1.67 mSec, providing every subscriber with the best possible rate at any given moment.

12.3.3 Separation of Voice and Data

- HDR's approach places data and voice on separate carriers allows better optimization for each, therefore higher capacity for each.
- Simplifies system software development and testing.
- Eases system operation and maintenance.
- Avoids difficult load-balancing tasks.
- Ideal data design complements existing and future voice networks.
- Dual-mode device can be integrated to provide optimum voice and data services.

12.3.4 Stand-alone System

- Decentralized option allows for stand-alone system deployment using off-the-shelf-IP backbone equipment.

13

What Does the Future Hold for Wireless Technologies?

All around the world, mobile phones are here to stay.

Mobile handset shipments are exploding globally, and by 2003 or 2004 (depending on whom you ask) handset makers will be shipping one billion units per year. In the United States alone, Cahners In-Stat Group, the Scottsdale, Ariz.-based market research firm, has forecast a 16.8% annual growth rate in new mobile phone users over the next five years. Globally, Cahners expects 1.87 billion mobile phone users by 2004.

13.1 COPS

Bell Labs, which is the R&D arm of Lucent Technologies and is also known as a long-time innovator in the telecommunications space announced a software breakthrough that enables global wireless roaming across all wireless networks. This includes wireless LANs using 802.11 technologies, CDMA2000, Universal Mobile Telecommunications Services (UMTS), and other high-speed data networks.

Bell Labs calls this software architecture COPS, which stands for Common Operations. COPS has been designed to facilitate access to voice/data services when subscribers are outside of their home networks, even when they are in different types of networks from their own. What this means is that a person using a phone operating on a Code Division Multiple Access (CDMA) network will be able to roam on a mobile network operating on the Global System for

Wireless Data Technologies. Vern A. Dubendorf
© 2003 John Wiley & Sons, Ltd ISBN: 0-470-84949-5

Mobile Communications (GSM). Cellular users will even be able to roam on to a WLAN supporting 802.11 standards.

The COPS architecture creates a so-called 'protocol gateway' which effectively translates data from networks employing disparate protocols into a single, common language. The result is that the various networks can maintain and use a single subscriber profile – including authentication, authorization, and location data.

13.2 Will Wireless LANS Hurt 3G?

The growing popularity and ubiquity of WLANs will likely cause wireless carriers to lose nearly a third of 3G revenue as more corporate users begin using WLANs to connect to the Internet and office networks. This is based on a recent report from the London-based market research firm Analysys.

Analysts say the ease of installing and using WLANs is making it an attractive alternative to mobile 3G. In contrast to the reported $650 billion spent worldwide by carriers to get ready for 3G, setting up a WLAN hotspot requires only an inexpensive base station, a broadband connection, and one of many interface cards using the 802.11b networking standard now available for your laptop, PDA, or smart phone.

Will WLANs supplant 3G? No. The two technologies use different radio frequencies and they are also targeting different markets. Where 3G is mostly phone-based and handles both voice and data, WLAN is a purely data-driven creation.

Allen Nogee is a senior analyst at Cahners In-Stat Group. Allen believes that rather than a threat, Wireless LAN technology will help introduce consumers to the wireless word and mobile commerce.

'WLAN technology allows customers to get accustomed to having wireless access, with no contracts to sign, and no commitment,' says Allen Nogee. Allen feels that after the initial experience, consumers are more likely to pay more for expanding their wireless reach beyond the limited range of WLAN.

14

4th Generation

As the roll-out of 3rd Generation (3G) wireless networks and services continue throughout Europe and the United States is just now beginning its use of CDMA2000 by Sprint PCS, work has already begun to define the next generation of wireless networks.

The move towards fourth generation (4G) wireless is made more difficult by the fact that a universal 3G standard has been adopted and deployed. Most industry experts agree that the future of wireless is one in which voice, video, multimedia and broadband data services traveling across multiple wireless air interfaces are meshed into one seamless network.

4G wireless networks will be recognized for:

- Seamless network of multiple air interfaces and protocols
- Improved spectral efficiency
- IP Based (probably IP v6)
- Higher data rates up to 100 Mbps

There are technologies that already exist which address many of the design challenges facing 4G developers. High performance processors today are able to meet the processing requirements of complex algorithms. The emergence of protocols such as RapidIO provide the means for high-speed, flexible and scalable data communications between processing elements.

There will be a marriage between high-performance processing elements and a switched-fabric interconnect creating a reconfigurable and scalable platform designed to overcome the challenges of 4G systems.

The technology to watch will be Wideband Orthogonal Frequency Division Multiplexing (W-OFDM).

Wireless Data Technologies. Vern A. Dubendorf
© 2003 John Wiley & Sons, Ltd ISBN: 0-470-84949-5

Wideband Orthogonal Frequency Division Multiplexing (W-OFDM) is a transmission design that provides for data to be encoded on multiple high-speed radio frequencies concurrently which allows for greater security along with increased amounts of data being sent as well as a more efficient use of bandwidth.

W-OFDM is the basis of the IEEE standard 802.11a, which is the foundation of the proposed IEEE standard 802.16.

W-OFDM is a patented technology in the United States under patent number 5 282 222 and in Canada under patent number 2 064 975.

W-OFDM technology is currently used in Wi-LAN's broadband wireless access systems and allows for the implementation of low power multipoint RF networks that minimize interference with adjacent networks. The results in reduced interference, which in turn enables independent channels to operate within the same band allowing multipoint networks and point-to-point backbone systems to be overlaid in the same frequency band.

From a technical standpoint, the 4G network, which is also being called "The Worldwide Network", will be more stable and intelligent then ever before. 4G is a superior technology when compared to the existing aging copper and aluminium local loop. 4th Generation Technology is also seen as a move from intelligence in the network or at the edges to intelligence everywhere. 4G is an all-IP based access and core with effective management of all types of QoS over IP, including handoff. Most likely 4G will beIPv6 based which is better adapted to mobile networks than IPv4 having adequate addressing capacity, multicast management, security mechanisms, QoS management, and mobility management.

Some of the Benefits of 4th Generation Technology include:

- Multiple functionalities in a single handset
  - Voice, bulk data transfer, image, short message, fax, Web surfing, video-conferencing/broadcasting and future applications, etc.
- Global roaming
- A single universal identification access number
- Seamless access, transparent billing, security
- Low cost in service and handset

Some of the Technical Challenges of 4th Generation Technology include:

- Resource allocation: multiplexing heterogeneous, bursty data traffic
- QoS guarantee for bandwidth and/or delay sensitive applications
- User channel scheduling: code assignment in CDMA
- Interoperability with 3G standards

- Ubiquitous deployment: indoor and outdoor cell coverage
- Convergence with backbone (wireline) networks

As we enter into the twenty-first century, the competitive landscape is undergoing radical change. Globalization of politics, economics, technology and communications appear unstoppable. While boundaries between countries and regions may be meaningful in political terms, with the advent of the Internet and now the globalization of information and communication technologies, these boundaries have all but disappeared. The ever-faster flow of information across the world has made people aware of the tastes, preferences, and lifestyles of citizens in other countries. Now with the availability of 3rd Generation Wireless Technologies and the work to produce the 4th Generation, being tied to a place to access this global information resource is going away. People will be free to travel and maintain constant connections to the world of information. Through this information flow we are all becoming global citizens and we only want quicker response, greater bandwidth and more capabilities such as Cisco's Advanced Voice and Integrated Data.

References

Schneier, B. (1996) *Applied Cryptography*, Second Edn, p. 758. John Wiley & Sons, Inc.

Siep, T.M. (2000) *An IEEE Guide: How to Find What You Need in the Bluetooth Spec.*, IEEE Press.

Telegraph Age, **November 1** and **November 15**, (1897). Marconi Telegraphy, *London Electrician*. (Reprint).

Brady, J.T. (1920) Talking by wireless as you travel by train or motor. *Boston Sunday Post*, **November 7**.

Wireless Data Technologies. Vern A. Dubendorf
© 2003 John Wiley & Sons, Ltd   ISBN: 0-470-84949-5

Acronyms and Abbreviations

ACK acknowledgment
ACL asynchronous connectionless link
ACO authenticated ciphering offset
AES advanced encryption standard
AG attachment gateway
AM_ADDR active member address
AR_ADDR access request address
ARIB Association of Radio Industries and Businesses
ARQ automatic repeat request
ARQN automatic repeat request negative
ASN.1 abstract syntax notation one
BB baseband
BCH Bose–Chaudhuri–Hocquenghem
BD_ADDR Bluetooth device address
BER bit error rate
BNEP Bluetooth network encapsulation protocol specification
BQA Bluetooth qualification administrator
BSIG Bluetooth special interest group
BT bandwidth time product (i.e. B*T)
CAC channel access code
CC call control
CDMA code division multiple access
CID channel identifier
CL connectionless
COD class of device
CODECS coder decoders

COF ciphering offset number
CRC cyclic redundancy check
CSMA/CD carrier sense multiple access with collision detection
CVSD continuous variable slope delta
DAC device access code
DC direct current
DCE data communication equipment
DCI default check initialization
DCID destination channel identifier
DH data-high rate
DIAC dedicated inquiry access code
DLC data link control
DLCI data link connection identifier
DLL data link layer
DM data-medium rate
DQPSK differential quadrature phase shift keying
DSAP destination address field
DTE data terminal equipment
DTMF dual tone multiple frequency
DUT device under test
DV data voice
ED energy detection
EIFS extended inter-frame space
ERTX expanded response timeout expired
ETC extreme test conditions
ETSI European Telecommunications Standards Institute

Wireless Data Technologies. Vern A. Dubendorf
© 2003 John Wiley & Sons, Ltd ISBN: 0-470-84949-5

FC frame control
FCC Federal
    Communications Commission
FCS frame check sequence
FEC forward error correction
FER frame error rate
FH frequency hopping
FHS frequency hop synchronization
FHSS frequency hopping
    spread spectrum
FIFO first in first out
FSK frequency shift keying
FW firmware
GAP generic access profile
GEOP generic object exchange profile
GFSK gaussian frequency shift keying
GIAC general inquiry access code
GM group management
HA host application software
    using Bluetooth
HC host controller
HCI host controller interface
HEC header error check
HID human interface device
HPC hand-held personal computer
HV high-quality voice
HW hardware
IAC inquiry access code
ICS implementation
    conformance statement
ICV integrity check value
ID identity or identifier
IDU interface data unit
IETF Internet Engineering Task Force
IP Internet protocol
IrDA Infrared Data Association
IrMC infrared mobile communications
ISDN integrated services
    digital networks
ISM industrial, scientific, medical
IUT implementation under test

IV initialization vector
L_CH logical channel
L2CA logical link control
    and adaption
L2CAP logical link control and
    adaption protocol
LAN local area network
LAP lower address part
LC link control
LCID local channel identifier
LCP link control protocol
LCSS link controller service signaling
LFSR linear feedback shift register
LIAC limited inquiry access code
LLC logical link control
LM link manager
LME layer management entity
LMP link manager protocol
Log PCM logarithmic pulse
    coded modulation
LP lower-layer protocol
LPO low-power oscillator
LSB least significant bit
M master or mandatory
MAC medium access control
MAPI messaging application
    procedure interface
MDF management-defined field
MIB management information base
MLME MAC sublayer
    management entity
MMI man–machine interface
MPDU MAC protocol data unit
MPT Ministry of Post and
    Telecommunications
MSB most significant bit
MSC message sequence chart
MSDU MAC service data unit
MTU maximum transmission unit
MUX multiplexing sublayer a sublayer
    of the L2CAP layer

NAK negative acknowledgment
NAP non-significant address part
NOP no operation
NTC normal test condition
O optional
OBEX object exchange protocol
OCF opcode command field
OGF opcode group field
OSI open systems interconnection
PAN personal area network
PAR project authorization request
PC personal computer
PCM pulse coded modulation
PCMCIA Personal Computer Memory
   Card International Association
PCS personal communications service
PDA personal digital assistant
PDU protocol data unit
PHT pseudo-hadamard transform
PHY physical layer
PICS protocol implementation
   conformance statement
PIN personal identification number
PLCP physical layer
   convergence procedure
PLME PHY layer management entity
PM_ADDR parked member address
PMD physical medium dependent
PN pseudo-random noise
PnP plug and play
POS personal operating space
POTS plain old telephone service
PPDU PHY protocol data unit
ppm part per million
PPP point-to-point protocol
PRBS pseudo random bit sequence
PRD program reference document
PRNG pseudo-random
   number generator
PS power save
PSM protocol/service multiplexer

PSTN public switched
   telephone network
QoS quality of service
RA receiver address
RAND random number
RF radio frequency
RFC request for comments
RFCOMM serial cable emulation
   protocol based on ETSI TS 07.10
RSSI receiver signal
   strength indication
RTS request to send
RTX response timeout expired
RX receive or receiver
S slave
SA source address
SABM set asynchronous
   balanced mode
SAP service access point
SAR segmentation and reassembly
SC scan period
SCID source channel identifier
SCO synchronous connection-oriented
SD service discovery
SDDB service discovery database
SDL specification and
   description language
SDP service discovery protocol
SDU service data unit
SEQN sequential numbering scheme
SFD start frame delimiter
SIFS short inter-frame space
SIG special interest group
SLRC station long retry count
SME station management entity
SQ signal quality
SR scan repetition
SRC short retry count
SRES signed response
SS supplementary services
SSAP source address field

SSI signal strength indication
SSRC station short retry count
SUT system under test
SW software
TA transmitter address
TAE terminal adapter equipment
TBD to be defined
TBTT target beacon transmission
  time
TC test control layer for the
  test interface
TCI test control interface
TCP transmission control protocol
TCP/IP transport control
  protocol/Internet protocol
TCS telephony control protocol
  specification
TDD time division duplex
TDMA time division multiple access
TS technical specification
TSF timing synchronization function
TTP tiny transport protocol

TX transmit or transmitter
TXE transmit enable
UA user asynchronous
UAP upper address part
UART universal asynchronous receiver
  transmitter
UC user control
UDP user datagram protocol
UDP/IP user datagram
  protocol/Internet protocol
UI user isochronous
URL uniform resource locator
US user synchronous
USB universal serial bus
UT upper tester
UUID universally unique identifier
w.r.t. with respect to
WAN wide area network
WAP wireless application protocol
WLAN wireless local area network
WPAN wireless personal area network
WUG wireless user group

Glossary

ad hoc network	A network typically created in a spontaneous manner. The principal characteristic of an *ad hoc* network is its limited temporal and spatial extent.
Asynchronous Connectionless (ACL) link	The ACL link is a point-to-multipoint link between the master and all the slaves participating on the piconet. In the slots not reserved for the SCO link(s), the master can establish an ACL link on a per-slot basis to any slave, including the slave(s) already engaged in an SCO link.
Attachment Gateway (AG)	The attachment gateway is a communications node with at least two communication interfaces, one of which is a Bluetooth interface and one of which is an interface to another network. An attachment gateway is used to attach a Bluetooth WPAN to the other network. In particular, an 802 LAN attachment gateway attaches a Bluetooth WPAN to an 802 LAN, while a PSTN attachment gateway attaches a Bluetooth WPAN to the PSTN network.
Authenticated device	A Bluetooth device whose identity has been verified during the lifetime of the current link, based on the authentication procedure.
Authentication	A generic procedure based on LMP-authentication if a link key exists or on LMP-pairing if no link key exists.

Wireless Data Technologies.   Vern A. Dubendorf
© 2003 John Wiley & Sons, Ltd   ISBN: 0-470-84949-5

Authorization	A procedure where a user of a Bluetooth device grants a specific (remote) Bluetooth device access to a specific service. Authorization implies that the identity of the remote device can be verified through authentication.
Authorize	The act of granting a specific Bluetooth device access to a specific service. It may be based upon user confirmation, or the existence of a trusted relationship.
Bluetooth baseband	The Bluetooth baseband specifies the medium access and physical layers procedures to support the exchange of real-time voice, data information streams, and *ad hoc* networking between Bluetooth units.
Bluetooth channel	A Channel that is divided into time slots in which each slot corresponds to an RF hop frequency. Consecutive hops correspond to different RF hop frequencies and occur at a nominal hop rate of 1600 hops/s. These consecutive hops follow a pseudo-random hopping sequence, hopping through either a 79 or a 23 RF channel set.
Bluetooth HCI	The Host Controller Interface provides a command interface to the baseband controller and link manager and access to hardware status and control registers. This interface provides a uniform method of accessing the Bluetooth baseband capabilities.
Bluetooth host	Bluetooth Host is a computing device, peripheral, cellular telephone, 802 LAN attachment gateway, PSTN attachment gateway, etc. A Bluetooth Host attached to a Bluetooth unit may communicate with other Bluetooth Hosts attached to their Bluetooth units as well.

Bluetooth unit	Bluetooth Unit is voice/data circuit equipment for a short-range wireless communication link. It allows voice and data communications between Bluetooth Hosts.
Bluetooth	Bluetooth is a wireless communication link, operating in the unlicensed ISM band at 2.4 GHz using a frequency hopping transceiver. It allows real-time voice and data communications between Bluetooth Hosts. The link protocol is based on time slots.
Bond	A relation between two Bluetooth devices defined by creating, exchanging, and storing a common link key. The bond is created through the bonding or LMP-pairing procedures.
Bonding	A dedicated procedure for performing the first authentication, where a common link key is created and stored for future use.
Channel establishment	A procedure for establishing a channel on L2CAP level.
Channel	A logical connection on L2CAP level between two devices serving a single application or higher layer protocol.
Connect (to service)	The establishment of a connection to a service. If not already done, this includes establishment of a physical link, link and channel as well.
Connectable device	A Bluetooth device in range that will respond to a page.
Connecting	A phase in the communication between devices when a connection between them is being established. (Connecting phase follows after the link establishment phase is completed.)
Connection establishment	A procedure for creating a connection mapped onto a channel.
Connection	A connection between two peer applications or higher layer protocols mapped onto a channel.

Coverage area	The area where two Bluetooth units can exchange messages with acceptable quality and performance.
Creation of a secure connection	A procedure of establishing a connection, including authentication and encryption.
Creation of a trusted relationship	A procedure where the remote device is marked as a trusted device. This includes storing a common link key for future authentication and pairing (if the link key is not available).
Isochronous user channel	Channel used for time bounded information, like compressed audio (ACL link).
Kilobyte (kb)	1000 bytes.
Link establishment	A procedure for establishing a link on LMP level. A link is established when both devices have agreed that LMP setup is completed.
Link	Shorthand for an ACL link.
LMP-authentication	A Link Manager Protocol level procedure for verifying the identity of a remote device. The procedure is based on a challenge–response mechanism using a random number, a secret key, and the BD_ADDR of the non-initiating device. The secret key used can be a previously exchanged link key.
LMP-pairing	A procedure that authenticates two devices, based on a PIN, and subsequently creates a common link key that can be used as a basis for a trusted relationship or a (single) secure connection. The procedure consists of the steps: creation of an initialization key (based on a random number and a PIN), creation and exchange of a common link key and LMP-authentication based on the common link key.
Logical channel	The different types of channels on a Physical Link.

Mode	A set of directives that defines how a device will respond to certain events.
Name discovery	A procedure for retrieving the user-friendly name (the Bluetooth device name) of a connectable device.
Packet	Format of aggregated bits that can be transmitted in one, three, or five time slots.
PAGE	A baseband state where a device transmits page trains, and processes any eventual responses to the page trains; terms written with capital letters refer to states.
Page	The transmission by a device of page trains containing the Device Access Code of the device to which the physical link is requested.
Page scan	The listening by a device for page trains containing its own Device Access Code.
PAGE_SCAN	A baseband state where a device listens for page trains.
paging	A Bluetooth unit transmits paging messages in order to set up a communication link to another Bluetooth unit that is active within the coverage area.
Paired device	A Bluetooth device with which a link key has been exchanged (either before connection establishment was requested or during connecting phase).
Physical channel	Synchronized RF hopping sequence in a piconet.
Physical link	A Baseband-level connection between two devices established using paging. A physical link comprises a sequence of transmission slots on a physical channel alternating between master and slave transmission slots.
Piconet	In the Bluetooth system, the channel is shared among several Bluetooth units. The units sharing a common channel constitute a piconet.

POS	A Personal Operating Space (POS) is the space about a person or object that typically extends up to 10 meters in all directions and envelops the person whether stationary or in motion.
Pre-paired device	A Bluetooth device with which a link key was exchanged, and the link key is stored, before link establishment.
Radio	Communication without the use of wires other than the aerial; the ether and ground taking the place of wires.
RFCOMM client	An RFCOMM client is an application that requests a connection to another application (RFCOMM server).
RFCOMM initiator	The device initiating the RFCOMM session, i.e. setting up RFCOMM channel on L2CAP and starting RFCOMM multiplexing with the Set Asynchronous Balanced Mode (SABM) command on Data Link Connection Identifier (DLCI) 0 (zero).
RFCOMM server channel	This is a subfield of the TS 07.10 DLCI number. This abstraction is used to allow both server and client applications to reside on both sides of an RFCOMM session.
RFCOMM server	An RFCOMM server is an application that awaits a connection from an RFCOMM client on another device.
Scatternet	Two or more piconets co-located in the same area with inter-piconet communication.
SCO link	See Synchronous Connection-Oriented (SCO) link.
SDL (Specification and Description Language)	A modern, high-level programming language. It is object-oriented, formal, textual, and graphical. SDL is intended for the description of complex, event-driven, real-time, and communication systems.

Service discovery

Procedures for querying and browsing for services offered by or through another Bluetooth device.

Silent device

A Bluetooth device appears as silent to a remote device if it does not respond to inquiries made by the remote device. A device may be silent due to being non-discoverable or due to baseband congestion while being discovered.

Synchronous Connection-Oriented (SCO) link

The SCO link is a point-to-point link between a master and a single slave in the piconet. The master maintains the SCO link by using reserved slots at regular intervals.

TCP/IP

Transmission Control Protocol/Internet Protocol. TCP/IP is commonly used over the Internet wired computer network. The TCP/IP suite contains different transmission facilities such as FTP (File Transfer Protocol), SMTP (Simple Mail Transport Protocol), Telnet (Remote terminal protocol), and NNTP (Net News Transfer Protocol).

Time slot

The physical channel is divided into $625\,\mu s$ long time slots.

TNC

Terminal Node Controller. A TNC contains a modem, a computer processor (CPU), and the associated circuitry required to convert communications between your computer (RS-232) and the packet radio protocol in use. A TNC assembles a packet from data received from the computer, computes an error check (CRC) for the packet, modulates it into audio frequencies, and puts out appropriate signals to transmit the packet over the connected radio. It also reverses the process, translating the audio that the connected radio receives into a byte stream that is then sent to the computer. Most amateurs currently use

	1200 bps (bits per second) for local VHF and UHF packet, and 300 bps for a longer distance, lower bandwidth HF communication. Higher speeds are available for use in the VHF, UHF, and especially microwave region, but they often require special (not plug-and-play) hardware and drivers.
Trusted device	A paired device that is explicitly marked as trusted.
Trusting	The marking of a paired device as trusted. Trust marking can be done by the user, or automatically by the device after a successful pairing.
Unknown device	A Bluetooth device for which no information (BD_ADDR, link key or other) is stored.
Un-paired device	A Bluetooth device for which there was no exchanged link key available before connection establishment was requested.
Wireless	Communication without the use of wires other than the aerial; the ether and ground taking the place of wires.
WPAN	The term WPAN refers specifically to a wireless personal area network as outlined in the IEEE 802.15 standard.

Index

Wireless Data Technologies.   Vern A. Dubendorf
© 2003 John Wiley & Sons, Ltd   ISBN: 0-470-84949-5